FROM HANDWRITING TO FOOTPRINTING

From Handwriting to Footprinting

Text and Heritage
in the Age of Climate Crisis

Anne Baillot

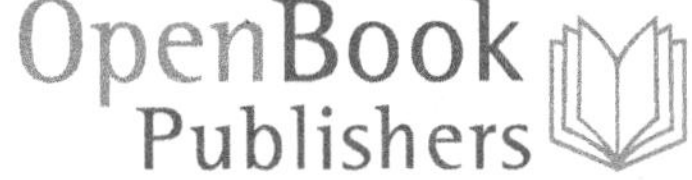

https://www.openbookpublishers.com

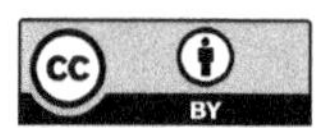

ISBN Paperback: 978-1-80511-087-3
ISBN Hardback: 978-1-80511-088-0
ISBN Digital (PDF): 978-1-80511-089-7
DOI: 10.11647/OBP.0355

Cover image:
Yarn texture by Tim Mossholder (2023), https://unsplash.com/photos/AQjSAPNLjGI

Cover design: Jeevanjot Kaur Nagpal

Contents

Introduction

In his famous lecture entitled *Encyklopädie und Methodologie der philologis-chen Wissenschaften*, classical philologist August Boeckh frames achieve-ments of the philological sciences as "die Erkenntnis des Erkannten", the knowledge of that which is known.[1] This definition of text-based scholarship is not as tautological as it may seem. It strives to embrace fundamental self-reflective efforts. Philologists — those who love texts — supply and interpret texts while they question the way in which they approach texts. In the same way, digital philology provides and inter-prets texts and reflects on the hermeneutical principles it follows. It also interrogates how digital media impacts textual transmission.

While text-based activities have remained essentially the same since Boeckh's time, including archiving, editing, commenting, publishing, and critiquing, the media shifts at play since the 20th century have opened new horizons and provided opportunities to engage productively with older cultural textual practices.

The advent of digital tools and resources has changed my own schol-arly endeavours in a radical way. From a solitary and oftentimes repetitive activity, whose output would concern only a handful of the like-minded, it became a collaborative endeavour in which mechanical tasks were con-ducted by a machine, an undertaking in which the whole world could potentially take part. It loosened the boundaries of scholarship and made them more permeable to adjacent fields, in particular cultural heritage.

While I have learned a lot both from this development of textual stud-ies through digital media, and from the manner in which it tightens the connections between philology and heritage institutions such as archives and libraries, this collaborative perspective is, to this day, still by no means predominant. Admittedly, it requires one to let go of established forms

---

1   Only one edition of the *Encyklopädie* manuscript currently exists. It was procured by Boeckh disciple Ernst Bratuschek in 1877 and presents all the available textual elements as one text, while the manuscript actually consists in over 20 layers corresponding to the updates that Boeckh made to his lecture script over time. Procuring a dynamic digital edition of this manuscript remains a desideratum. See Bratuschek, *Encyklopädie* [26], Horstmann, *Erkenntnis des Erkannten* [70], Baillot et al., *Neue Perspektiven* [15].

    https://doi.org/10.11647/OBP.0355.05

of authority, and with them, of power. Nonetheless, I am convinced that this approach is the only one that would make it possible to stay true to Boeckh's principle and, more generally, to root the social relevance of philology in the current context, which is one of digitisation, but also of the technical and cultural divide between the Global North and South, and of the world-wide climatic threat. Textual scholars and digital humanists will easily identify references to well-known theories in some of the following pages. Heritage professionals will also find attempts to popularise some of their advances for a non-professional audience. My writing is rooted in my personal experience as a scholar; it also reflects on a variety of theoretical and technological advances that have yet to benefit a wide user and reader community. It is my goal to address this community.

In this book, I discuss mechanisms that facilitate access to text based on my own scholarly experience of the last twenty-five years. I do so with a focus on early modern literary texts as an indicative part of cultural heritage. I present three core cultural and scholarly text-related practices — archiving, editing, and publishing processes — to frame them in a digital context, and ultimately shed light on the environmental impact of today's preservation and dissemination options for textual material. My goal is to show how theoretical and practical leverage can be gained from existing approaches in order to tackle the ecological and, with that, economical and social challenges we now face. I examine ways of preserving text and disseminating it to a wider audience from a historical, digital, and environmental perspective, and aim to foster reflections on today's archiving and publishing practices in terms of their sustainability in the context of the climate crisis.

Digital infrastructures have had a transformative influence on archiving and publishing ecosystems. Forms of media shifted, technical constraints appeared, costs rose, and reputation mechanisms took on new dimensions. I will argue, however, that modes of operation have not changed radically with the media transformations introduced during the 20th century. But, more recently, digital media have managed to move some philological lines, and the climate crisis calls for an even more enormous reassessment of practices of textuality. In the first two parts of the book, I move from general historical and theoretical remarks on archiving in Chapter 1 and publishing in Chapter 2 to their digitisation.

In the final part, Chapter 3, I explore the impact of the climate crisis on current archival and editorial practices.

Chapter 1 is dedicated to archiving strategies and their significance for our relationship to text and to the value of text. After a general presentation of archiving processes and the institutionalisation of archives in western societies, I go on to highlight the specificities of digital archiving. Trying to assess continuity lines and rupture points between traditional (analog) archiving and digital archiving, the argument focuses in this section on the materiality of both processes. Recording strategies are then presented as major leverages. The last section in Chapter 1 reflects on a more theoretical level about the contradictions inherent in text archiving as a balance between preservation and destruction of material.

Chapter 2 addresses publishing practices and strengthens the focus on literary texts. I present early modern publication processes as transformative events, and assess the impact of digitisation. The first section sheds light on the process of transforming a text into a book and ultimately part of an œuvre through edition and publication, examining authorship issues along the way. The relationships between writers and publishers play a key role here. I present a case study on two major German writers of the first half of the 19th century, Goethe and Tieck, in order to give a better sense of the pragmatics of authorship negotiation. The second section presents a panorama of current digitisation processes (scanning, automated text recognition, annotation, visualisation) and the way in which they multiply the forms of text representation available to readers. The closing section reflects on the implementation and social significance of Open Access both for archiving and for publishing.

Chapter 3 addresses the environmental footprint of digital archiving and digital publication. It suggests new approaches to the question of access to text in a context of digital sobriety, being frugal of one's use of diverse technologies. In the first section, I frame the overall environmental footprint of access to text in a digital context, underlining the difficulty of measuring it precisely and making informed decisions throughout the process of archiving or publishing. From a more speculative perspective, I then envision what sustainable access to text could look like in the future. The last section focuses more specifically on the production, dissemination, and storing of the book you are currently reading. It aims to identify the type of leverage that is necessary to offer a transformative, less resource-intensive approach to (digital) text.

I conceived these analyses as a journey through text. It begins in the following pages with old family papers found in a drawer and explores the different ways of preserving them and making them available for more readers. I argue that every reader is key in the transmission process: from the moment you are given access, you become one who can also give access. In this process, a new balance is established between the growth of digital transmission options, the almost magical immensity of content they provide, and their impact on the physical integrity of our planet. My argument is a material one: it is one of love for texts, especially literary texts, as one of the richest means of human expression — but one that pleads for a better embedding of textual activity in the complex materiality of our world.

1. Archiving text

Text presents itself to us in everyday life in a variety of forms, to the point that it has become key to more or less all social, political, economical, and cultural transactions in north-western cultures. Text is embedded in the immediacy of our environment. Yet not all textual material that surrounds us is a product of immediacy. I would like to begin here by looking at an approach that connects us to the past of textuality. What happens when you consider an old text?

Old papers have a fascinating quality that unfolds when you come into close contact with them. They can be kept in bundles or simply present themselves as loose sheets. Some might seem easy to decipher, while others are virtually unreadable. Sometimes they are composed of neatly numbered pages, sometimes of incoherent fragments. None of that makes much difference. What matters is that, most of the time, something happens to those whose path crosses old papers, whether the encounter happens on purpose or by chance.

The way in which old papers present themselves, their materiality, creates a distance. The paper differs from what one is used to flipping through in everyday life; it has aged. The shape of the letters or ciphers is also different, as are the ink and pen that are used. Nevertheless, it is a product of paper and writing and in that sense it is familiar. This mixture of familiarity and distance is captivating: it feels as though it would be possible to unlock the mysteries old papers contain — and at the same time it is clear that this can never be fully achieved.

Old papers transport you back in time while simultaneously anchoring you in the reality of your present time and space. Even if you manage to decipher their content, understanding fully what they are about will require much more effort than simply reconstituting the wording. After transcribing the letters and forming words from them, you need to understand where the papers came from, who the writers were, what their intention might have been, who or what they were alluding to, but also, why these specific papers have been preserved through time and finally landed into your hands.

    https://doi.org/10.11647/OBP.0355.01

Preservation is far from a general rule when it comes to inheriting the past. Old papers rather tend to disappear or disintegrate as time goes by. Paper technology is remarkable, but it resists water and fire rather poorly. Sometimes even air can induce material decay, especially if paper and air react chemically with aggressive ink types. And it is not only physical elements that accelerate decomposition: rodents too have a fair share of responsibility in the poor preservation of old papers.

And yet, when you find them, there they are, here and now. There is something compelling in the presence of old papers, as if they were a gift of time past. The specific piece of paper you are holding has resisted severe weather, wars, mice, neglect and mould, and it has done so for decades or centuries, as if it were all just to reach you, today. You might be afraid that you will damage it. The paper is probably becoming tattered, especially on the edges, and might be breaking up into smaller pieces where it was folded. But, for all its frailty, it has proved its extreme resistance simply by still being there.

While some old papers — family papers randomly found in a drawer, for instance — might not seem to obey a preservation logic destined to lead them to you intentionally, for a very long time there have been institutionalised forms of preservation, dedicated to a range of documents. Intentional preservation, with the goal of making items available at a later time, is in fact an accurate definition of what archives are and what they do.

Archives are, by and large, organised old papers. An archive consists of a coherent ensemble of old papers, as in "The Archive of the Paris Police". The word "archives" also designates institutions and/or the buildings hosting them, which take care of such organised old papers. The singular "archive" or its plural can be used to designate any one of them: the collection of papers, the institution, or the place where the institution is. The function of archives is to store, record, and present old papers from various times and places, and with various subjects. Whilst the content of an archive is not always restricted to handwritten papers, these generally make up the core of their stock.

The existence of early archives (and I will get to a historical overview of archival processes in section 1.1) suggests that archival intentions have existed in human cultures for quite some time. For centuries, there have been reasons for mankind to make an effort to preserve written traces of their political, economical, legal and cultural achievements. Since archiving had a purpose, and presumably followed a strategy, anyone handling old papers today can legitimately wonder why they had been archived and, more specifically, why one is, in this particular place and at this particular moment, holding these remnants of time past. Getting in close contact with old papers also involves acknowledging your responsibility to pursue the transmission momentum beyond your own personal experience — that special moment when you come to see, touch, unpack, unfold, present to the light and hold old papers.

Preservation does not generally happen from one day to the next. Archiving is a process. It is the process by which, on the one hand, contemporary material is turned into archival content, being stored, recorded, and made available for consultation in an institutionalised manner. But, on the other hand, it reflects the ongoing effort that is necessary to keep old papers remembered: transmission is never achieved once and for all, but needs to be actively pursued in order for the archive to continue to exist. In that sense, anyone consulting archived material is part of the archiving process. At the exact moment when you are consulting the archive, you become a recipient and a transmitter. Archiving is not exclusively the mission of institutions like archives: it engages each member of a society that is concerned with the past and its memory.

To begin this journey to and through text, this chapter will examine archiving as an old cultural technique — one that is strongly rooted in north-western societies. As a first step, I will present a historical approach to archiving techniques and go from there to the specificities of digital archiving, in order to highlight differences but, in particular, structural similarities between physical and digital archives. The second section is more theoretical: I propose a conceptualisation of approaches to archival material as a fleeting trace of things past. This chapter aims to show how past texts are embedded in our present lives, both materially and symbolically, and how we can draw from this to make our present become tomorrow's lively past.

1.1   What archives do

Archives and archiving relate to the material world: they concern the documents you find in a family home, the postcard collection you might purchase at a flea market, and the notary acts recorded throughout the history of your city, as well as medieval copies of illuminations. This meaning of "archive" is the primary reference of the word, as applied to archives as institutions, and archiving as a process related to this way of collecting and recording documents. The same words have a slightly different meaning in the digital context. I will first provide some historical background on the institutionalisation of archiving processes in the analog world, and then move on to the specific challenges of digital archiving.

1.1.1   Institutionalisation of archiving

The word archive itself comes from the Ancient Greek word αρχη, which generations of Hellenists have learned to identify as "the word that means either the beginning or the commandment". The point is that it actually means both, relating to something like a seminal order of things.[1]

It is quite clear how the word "archive" relates to the second meaning of the Greek word, that of commandment: archiving provides order. For each archive, there is a rule, or, more accurately, a set of rules, that is to be respected throughout the preparation of the material in order to achieve a satisfactory archiving process — one that will make it possible to preserve, stock, record, and make available its contents for a later consultation.

Recording strategies only work if the same logic is used to structure all the material that belongs together: this is the requirement that enables the organisation and listing of the documents that make up the archive in a manner that will make it possible to eventually find them again when one looks for them. Alphabetical order, for instance, is a simple way of arranging material. But you first have to decide what the alphabetical order applies to: the author's name (then what about unknown authors?), the document's title (provided there is one), or maybe simply the beginning of the document's text (easier to identify if the text is linear and running). And then you have to decide how to organise the material in the space

---

1   Derrida opens his small opus *Mal d'archive* with exactly these etymological considerations ("Ne commençons pas au commencement, ni même à l'archive. Mais au mot "archive" — et par l'archive d'un mot si familier. *Arkhè*, rappelons-nous, nomme à la fois le *commencement* et le *commandement*. Ce nom coordonne apparemment deux principes en un: le principe selon la nature ou l'histoire, *là où* les choses *commencent* — principe physique, historique ou ontologique —, mais aussi le principe selon la loi, *là où* des hommes et des dieux *commandent*.") [42], p. 11.

at your disposal: will you, for instance, fill first all the upper shelves on one wall, then the row below, or use separators such as the supports between shelves to switch level to proceed with the rest of the alphabet? Also, you might want to consider the fact that your collection could grow at a later point, which would require a shelving system that allows for enlargement.

Obviously, even for a very generic archiving task, there are many ways of organising documentary material, and what may seem intuitive to one person might not be self-explanatory at all to another. Organising material requires an explicit set of rules, decisions on a variety of sorting steps, and clear, accessible information about these decisions and rules.[2]

No wonder then that archives have contributed to the development of standardised recording methods across time. General archiving principles make it easier to consult different archives. Archival logic follows similar, general rules, regulated and reflected on by what has become a field of knowledge of its own: archival sciences as a branch of information sciences.[3]

Even if you have only ever had to deal with a small set of documents, you are certainly aware that a lack of systematic organisation inevitably leads to much time-wasting looking for information or documents you know you have but cannot remember where they are. Optimising organising and recording strategies is at the core of archiving processes, as archives, by definition, have to deal with large, and generally growing, amounts of material.

In Ancient Athens, the creation of the first institutional archive, the Metrôon, was initiated in order for citizens to be able to consult any law that was passed. This is the first known systematic, institutionalised preservation and accessibility device, including a dedicated building conceived to house one copy of each law that was ever passed.[4] Laws were

---

2   See Petra Gehring, "Archivprobleme", in *Handbuch Archiv* [58], pp. 17-18.

3   Anne Gilliland provides a rich overview of the history of archival sciences in Chapter 1 (entitled "Archival and Recordkeeping Traditions in the Multiverse and their Importance for Researching Situations and Situating Research") of *Archival Multiverse* [61], pp. 31-73; see especially p. 43 for a definition.

4   This does not mean that there were no other archival endeavours before the creation of the Athenian Metrôon. In the volume edited by Marie Brosius (*Ancient Archives and Archival Traditions*), the focus on the Athenian model of the 5th century B.C. is limited to one contribution, while many chapters examine other contexts. The contribution dealing with Greek archives by John K. Davies suggests that the mass of documents

transcribed on stone, not paper, and only full citizens were authorised to consult them. But it was already structured along the lines of what archives would be in the following centuries in European societies: a building dedicated to the systematic preservation of information pertaining to public matters, and a place where they could be consulted. In the case of the Metrôon, the information was legal in nature, and it could serve political purposes. Being able to refer precisely to pre-existing laws improved transparency in political and judicial matters. Instead of giving only an approximative account of a law, one could retrieve the actual wording.

The Metrôon building was conceived for the preservation of documents, which means that the format and storage conditions, as well as the fact that the stock would grow over time, were accounted for from the onset. Also, the preserved material did not consist of cultural artefacts, but documents, albeit carved on stone, relevant to the political well-being of the state, administrative material providing the history of the judicial branch of government.

These archives had a double function. First, they guaranteed the veracity of decisions taken in the past. People who needed to refer to these laws could do so confident of their accuracy, so that any suspicions about errors in the cited wording could be fact-checked. It was also a way to supply a precise remembrance of the known laws. People who might otherwise overlook them, not necessarily for manipulative reasons, but simply because their memory could not encompass all of the information inherited from previous generations (or even their own), would need some kind of memory extension that could provide access to this wide collection of all legal decisions made in the past.

Archives act as an ancillary service to human memory for the benefit of society at large. Archiving documentation connected to genocides like the Shoah, for instance, is strongly motivated by the wish for such an

archived in Athens was quite unique: "Argument has focused mainly on late fifth-century Athens. On the one hand, her headlong development, and her management of an Aegean empire, generated far more public documents than were ever cut in stone. [...] Nor was the practice purely Athenian. [...] Though no precise 'solution' to the problem is available, the general direction in Athens is clear. By the end of the fourth century BC at latest there was a reasonably well-organised public archive, located in the precinct of the Mother of the Gods, the Metrôon, wherein documents were lodged and could be found." (See [30], pp. 328–329). The main focus of Davies' analysis is the complex question of public access to the collection.

event never to happen again: archives provide details that people might end up forgetting, risking the denial of the event and its significance. World War Two is in many ways a turning point in public strategies of memory building in western societies, based on the theoretical work of the Frankfurt school of sociology, among others.[5]

More recent examples spring to mind, illustrating the relevance of the issue. During the invasion of Ukraine by Putin's Russia in the spring of 2022, alongside international manifestations of solidarity with Ukraine, there has been a noticeable effort worldwide to support the preservation of Ukrainian heritage.[6] The awareness that not only cultural, but also economical and political issues are at stake when cultural heritage is destroyed, has risen to the point that it is clear that its destruction would facilitate the negation of Ukraine's political existence as a nation. This sensibility is aligned with theoretical approaches that have developed since the 1970s, which shed light on issues that were previously chiefly the concern of archivists.

In his small book dedicated to archiving processes, *Mal d'archive*, French philosopher Jacques Derrida combines the notion that an archive would be an additional memory, or an extension to human memory, with the idea that it is the archiving process itself that generates the relevance of that which is archived. He presents archive as a memory supply ("prothèse ou technique hypomnésique") and as the generator of the event it records through the archival process ("archive archivante", "produit autant qu'elle enregistre"), but also the need for an infrastructure ("structure technique") backing both of these functions. Derrida postulates that archiving produces and records events: in his view, recording is that which produces the archive as such.[7]

---

5   See Horkheimer and Adorno, *Dialektik der Aufklärung* [69].

6   See, for instance, the SUCHO initiative for online content. Other initiatives tried to provide support for digitising what could be digitised, and protecting what could be protected. The symbolic and economic importance of cultural heritage preservation has been reported on, as far as this was possible, by journalists; see, for instance `https://www.nytimes.com/2022/04/30/world/europe/ukraine-scythia-gold-museum-russia.html?smid=tw-nytimes&smtyp=cur`.

7   See Derrida, *Mal d'archive* [42], p. 34: "Autre façon de dire que l'archive, comme impression, écriture, prothèse ou technique hypomnésique en général, ce n'est pas seulement le lieu de stockage et de conservation d'un contenu archivable *passé* qui existerait de toute façon, tel que, sans elle, on croit encore qu'il fut ou qu'il aura été. Non, la structure technique de l'archive *archivante* détermine aussi la structure du contenu *archivable* dans son surgissement même et dans son rapport à l'avenir. L'archivation produit autant qu'elle enregistre l'événement."

Derrida also provides a definition of recording ("consignation") that can be read as the affirmation of an identity that lies between recording and archiving.[8] Archival recording assembles elements according to one system, which thus generates unity. In these terms, the core principle of the archive is its system of recording.

Derrida's analysis reads like an abstract characterisation of archiving as a way of putting boxes and catalogue labels together and considering that they then constitute one unit. The specific space for assembling the archive is essential. As Derrida underlines: recording is a concrete, material process without which archives would not even exist.[9]

But it could also be read the other way round. If recording has an essential, transformative function in the archiving process, then maybe recording is sufficient to transform anything into an archive. Not so. You can find this out the hard way, as I did. When I was a young adult, I found some old family papers dating back to the late 19th century, and wanted to give them to the relevant local archives, which kept administrative papers recording regional history. What I brought were school papers and ledgers, but this was not in itself significant: I was sent back home with all of them altogether, because such material was only considered worth archiving if it was at least 150 years old. I have put the papers back in the cellar, where they still are to this day, waiting to be over 150 years old and deemed worth recording and professional archiving.

Obviously, if everything was to be considered valuable enough to be archived, the existing infrastructure dedicated to archiving would have to be much larger than it is. There would be whole buildings devoted to this: I imagine them to be an underground replica of actual cities, each one of them having a sort of "below the water part of the iceberg" underground mega-archive recording each city's story so far (a fantasy that probably unfolded in my imagination under the impression of Freud's depiction of Rome I mention at the end of section 1.2). In real life, and regardless of my personal dream of a mega-subterranean archive, archives have to

---

8   See Derrida, *Mal d'archive* [42], p. 14: "La consignation tend à coordonner un seul corpus, en un système ou une synchronie dans laquelle tous les éléments articulent l'unité d'une configuration idéale. [...] Le principe archontique de l'archive est aussi un principe de consignation."

9   The development of recordkeeping systematics has been central in the development of archival sciences since the 18th century, as accurately described by Gilliland in *Archival Multiverse* [61], especially pp. 38-39.

define priorities in order to achieve their missions. This can be stating that documents have to be at least 150 years old to be integrated to the archives, as was the case in my youthful experience; or deciding to collect only documents related to a specific topic, such as the legal focus they had in Athens; or being strictly dedicated to a particular political entity, such as a parliament, for instance.[10] This choice of a scope to be determined for each archival institution, which defines the extent of what this institution will preserve, record, and make available for consultation, is embraced in the concept of appraisal, championed by archivist Theodore Schellenberg in the United States in the 1950s.[11] From his perspective, the archivist's role is not simply to record documents, but also to decide which documents are to be kept and recorded.

Archives as institutions conceived to keep traces of the history of a state have developed their own purposes and practices alongside other cultural heritage institutions. While they sometimes converge in techniques and purposes, there are key differences between archives and libraries or museums. Archives preserve written documents that are, by definition, unique (or of which only few or, more likely, no other copies exist): in this, they are different from both museums and libraries. Museums are dedicated to artefacts and plastic works of art, while libraries provide single exemplars of books that were originally printed in several, oftentimes hundreds of copies, meaning that one would be likely to find the same book in another library. By definition, there is no automated reproduction of the content of an archive as opposed to a library, and the content consists of written material, in general handwritten, presented on paper as opposed to a museum. There are museums that display archival material as well as archives that preserve printed content. But there is a material difference that characterises the basic mission of each institution.

Another specific characteristic of archival structure concerns the manner in which unique documents are recorded. While alphabetical order and shelf structure, which I mentioned before, play a role in the organisation of boxes and their labels, modern archives have one central structur-

---

10 Similarly, the European Union provides an online archive of parliamentary minutes in the different official languages of the EU, see `https://historicalarchives.europarl.europa.eu/home.html`.

11 See Schellenberg, *The Appraisal of Modern Public Records*, [94]. See also the definition in the SAA *Dictionary of Archives Terminology*: `https://dictionary.archivists.org/entry/appraisal.html`.

ing principle — provenance.[12] The recording process of a document or a set of documents in archives is based on where they came from at the moment they were included in the archives' holdings. This principle is key to the structure of archives world-wide, not just the west, as it was exported in the wake of colonisation.

If all documents had always been archived on the basis of a mission-based, coordinated logic in, say, one political entity, this would be fairly straightforward. But the simple fact that political entities change over time already suggests that things are bound to be more complicated. We could imagine, for instance, that in the 19th century all documents to be preserved regarding the legal body of one nation were gathered in one place, and dispatched later on to regional archives for practical reasons: in that case, the inherent logic of the set of documents would remain untouched. But what really happened, especially in the late 19th and early 20th, is that a lucrative black market developed for manuscripts — often single documents or even pages, without their archival context — which led to the destruction of hitherto coherent collections. Additionally, both world wars saw forms of archival pillaging, or the displacements of documents away from combat zones, which led to the dismantling of collections and document loss.

Let me take an example here to get a better sense of the practical issues that the principle of provenance entails. I draw here on my interest in the collections of the *Staatsbibliothek zu Berlin*, which has a large manuscript department holding, among others, collections of handwritten documents pertaining to 19th-century intellectual and literary life, and in particular to Romanticism. There, I would happen to find manuscripts by the same author kept in different archival units. Indeed, manuscripts by a single author could be split between different boxes, or *Nachlässe*, while others were recorded as single manuscripts or *Autographen*. Following the principle of provenance, the documents were sorted depending on where the manuscript had come from when it was acquired and added to the stock.

For several decades, the acquisition policy of the institution has been to try and complete thematic archival stocks that were already substantial,

---

12   "Respect des fonds" in French, "Provenienz" in German. It was theorised as early as 1898 in the so-called *Dutch Manual* that served as a basis for archivists worldwide in the following decades. Gilliland provides an informative, albeit critical overview in [61], pp. 38-39.

and to document new additions as precisely as possible, especially their provenance. This impacts greatly on the collections connected to German Romanticism.

The "filling the gaps" strategy, especially in the context of manuscripts dealing with 19th-century intellectual life, owes a lot to manufacturer Ludwig Darmstaedter who donated his manuscript collection to the ancestor of the *Staatsbibliothek zu Berlin*, the Royal Library of Berlin. It also owes a lot to the fact that it was librarians — and not archivists — who determined the recording strategy for this collection and its later goal of "completion".

Darmstaedter lived at the end of the 19th century and was a great admirer of science and the history of science, a field in which he acquired many manuscripts. His private collection was structured according to scientific areas as they were defined at the end of the 19th century. What he eventually donated to the Royal Library was an extremely large collection of manuscripts of all kinds, documenting the many areas he considered relevant to the history of science in earlier centuries. Thus, although voluminous, the collection was not exhaustive, but rather focused on the then relevant fields of scholarship. But it was still large and encompassing, and its structure had a strongly systematic character.

This aspect was so pronounced that the documents that were acquired at a later point by the *Staatsbibliothek zu Berlin* were taken from their context of acquisition in order to integrate them into the Darmstaedter collection, with the goal of making it fulfil the mission the donor had originally assigned to the collection — much to the dismay of archivists. However, this structure has advantages, as Jutta Weber, former head archivist at the *Staatsbibliothek zu Berlin* argues, saying that dismantling some collections led to the construction of unequalled novel insights. Weber justifies *a posteriori* a strategy of destruction of an archival unit for the benefit, in particular, of exploiting correspondences and reconstructing historical networks. On the one hand the "context of collections of papers" is being destroyed, but, on the other hand, it facilitates the creation of a "cosmos", a whole universe.[13] This example encapsulates a virtual war between

---

13   See Weber, *Sternstunden eines Mäzens* [101], p. 49: "Natürlich wurden hier die Zusammenhänge von Nachlässen zerstört, aber was wurde gewonnen? Ein wissenschaftlicher Kosmos, der heute seinesgleichen sucht. Die Beziehungen von Wissenschaftlern zueinander wurden durch die gemeinsame Verwaltung ihrer Korrespondenz in einer Sammlung von Augen geführt."

two irreconcilable orders. The first made provenance the main principle of its organisation: it was to be dismantled in favour of the second one, the reconstruction of historical networks, which in this case gained the upper hand on the premise that it would provide greater advances in knowledge.

Today's archivists do not randomly dismantle acquisitions. On the contrary, they try to preserve provenance coherence. This means that, for instance, if the missing page of a manuscript is acquired separately at a later point, it will not be physically reintegrated at the "right" place in the order of pages because its provenance is different. It will most likely be presented as an autograph item in and of itself. The connection between the two archival units then needs to be explained for a reader to be able to put the pieces together again, and consult the manuscript as the whole it used to be.

A thorough documentation, laying out the content of all archival units, is required to make it possible for the reader to reunite elements that belong together — if not by way of provenance, then in terms of their context or topic — and to organise the consultation of these documents accordingly. This documentation takes the form of a catalogue or index, fittingly called a "finding aid" in archival contexts: they truly do help find the elements one is looking for. Considering collections as puzzles where time has separated pieces from one another, archives, while they may store the pieces in different boxes, provide finding aids, which indicate how to put them together to make the puzzle as complete as possible, if only for the duration of a single consultation.

Provenance logic has the intrinsic advantage of making shifts and movements between hosting institutions somewhat swifter. If we consider archival material that was displaced during World War Two, the political dimension of a change in the hosting institution and the conflict potential it involves, are not to be underestimated. Again, 19th-century manuscripts from the *Staatsbibliothek* offer a good illustration. During World War Two, part of the holdings dedicated to early 19th-century literary life was taken to the city of Merseburg in order to be kept safe from shelling: it was never recovered. Another part of the stock was taken to Krakow, and it is still there today. For decades now — at the very least since the end of the Cold War — there have been ongoing discussions between Germany and Poland concerning these manuscripts, Berlin still considering them theirs, as does Krakow.[14]

Part of the "Krakow deposit" transferred from the Berlin archives is the famous Varnhagen collection, a treasure for the study of German Romanticism. Just as with Darmstaedter's corpus, it had all started with an enthusiastic collector, in this case Karl August Varnhagen von Ense. Varnhagen first collected the papers of his late wife, Salonnière Rahel Levin, as well as her and his own correspondence with key actors of German Romanticism. He then extended his collection to further manuscripts documenting early 19th-century German intellectual life. Upon his death, his niece Ludmilla Assing inherited the collection, and eventually donated it to the Royal Library — the future *Staatsbibliothek*.[15]

The challenges of recording the Varnhagen collection extend to four areas: first, the physical place of deposit of the material artefacts; second, access to the catalogue that provides information on the content of the collection; third, the original geographical provenance of the stock; fourth, (political) relationships between two institutions, here the *Staatsbibliothek* on the one hand and the Jagiellonian University Library in Krakow on the other, or, more generally expressed, the countries who claim to be the legitimate hosts for the collection (the locations of the current hosting institution and the prior hosting institution). The preserved manuscripts are the bearers of all of these layers of ownership.

This type of challenges has a very concrete material impact on the manuscripts since each hosting institution has its own stamps and signs with which it marks — in the most literal way — the collections that are its own. Each has its way of numbering pages, too. The type of stamp or mark you can find on the manuscript (in some cases, on each page of a manuscript) also provides information on the period of time in which an institution was its owner. All in all, the integration of a handwritten document into a host institution makes the political, cultural, and ideological agenda of each collection visible on every document, sometimes on every page.

---

14   The presentation of the collection in the German general catalogue maintains a tactful ambiguity in that regard, see `https://kalliope-verbund.info/de/ead?ead.id=DE-611-BF-24146`.

15   Nikolaus Gatter inspects provenance issues regarding this collection in *Lebensbilder* [57] and in *Gift, geradezu Gift* [56].

The difficulties that arose in the wake of World War Two for the two Berlin collections I have mentioned show that provenance is not just a recording principle, but an authority issue, even if for different reasons. Discussions regarding the Varnhagen collection are still ongoing to this day: the case is not closed. Almost more dramatically even, the fate of the Darmstaedter collection epitomises German political history. It was transferred to Franconia during World War Two and successfully preserved, but when it was returned, the corresponding record cards, that is, the finding aid documenting the content of the collection, were sent to East Berlin, while the collection itself went to West Berlin.[16] The collection had to wait until German reunification to be merged again under one roof.

Recording, to speak in Derrida's terms, is not solely a question of organisation and structure ("consignation" as a "principe archontique"), but is essential in addressing the existence of archival material, its origin, and its integration into a specific archival context. While Derrida's stance is primarily theoretical, the examples I have given show that this theory is rooted in historical contexts: any archive is as much part of a historical setting as of the context of its current consultation. Any archive lives in at least two temporalities, and keeps recalling the multiple spaces and times in which it has been collected, kept, and eventually recorded and catalogued.

Ancient Athens had the advantage of being composed of little more than a city, which meant that taking decisions regarding the state and its way of working was materially not as challenging as it had become in early modern Europe. The Metrôon was one address, in the city centre, close to where all other central political institutions were located. In 18th-century Europe, political entities, such as states, had gained a relative political stability in terms of borders or the form of government, but they were much more complex, geographically, socially, and economically widespread structures than was the case in Athens in the 5th century B.C.. And so too was the political project behind their archiving strategies.

As my examples have already suggested, archiving strategies are strongly connected to political goals, and especially to the construction of national narratives. This involves not only administrative material,

16 See the article by Gabriele Spitzer in the exhibition catalogue *Sternstunden eines Mäzens* [101], here p. 28.

such as police, court, or parliamentary documents that are deliberately archived by a political authority, but also cultural artefacts. The shift from an administrative to an (also) cultural function was performed at different moments depending on the country. Even in a late-bloomer nation like Germany, perhaps even more so there, the choices that were made to conceive and implement national unity relied strongly on strategies in which particular writers were set up as symbols of national culture, not least through the creation of dedicated archival spaces.

One could hardly find a better example than Goethe to illustrate this point. Goethe lived to be 83 years old — long enough to look back on his own life and conceive not only several late, more or less final, editions of his works, but also an archive of his personal papers. In the last five years of his life, he pursued several projects in parallel (including the publication of his correspondence with Schiller, and completing *Faust II*), and he also set his secretaries onto the task of sorting and organising an archive of his life's achievements and work. He kept all these documents and partly reviewed them himself. Writing about it in his correspondence, he mentions estrangement ("I can't remember having ever given this topic any kind of attention"), the enormity of the task ("this is never ending", "it takes up so much space"), but eventually concludes he has to fulfil this mission for the sake of posterity, of the nation.[17]

I will not comment further on the obviously high regard Goethe had for himself and his work. What he is articulating here is interesting for other reasons. First, he is driven by the notion of the nation's interest. The nation is presented here as a *Kulturnation* by essence. What is relevant for the cultural elite is, in his view, relevant for the whole nation: politicising society at large means including cultural elements that, strictly speaking, actually concern only a fragment of the population (the well-read elite). Second, Goethe suggests that such a task is too much for a single person, even one with many diligent aids to assist and no pressing money issues like himself. The next logical step would be to say that "the nation", that is, political decision makers, should take the matter into their own hands.

Goethe was very well aware of the power of archiving. Controlling the way his work and life would be archived gave him a unique *post-mortem* hold over the reception of his work. This involved, obviously, sorting

---

17   See Baillot, *Moi, solitaire, tel Merlin* [11].

out which documents were to be kept and which were to be eliminated. But it also concerned the way in which they should be classified and presented. By presiding over the constitution of an archive of his life and work, Goethe determined the definition of his œuvre such as he intended others to see it,[18] and in the way he wanted to see the tradition established. After his death, the papers went over to his grandson, who made the great Duchess of Weimar sole heiress. She is the one who initiated the erection of a dedicated building, the current Goethe-und-Schiller-Archiv in Weimar. The Schiller papers, as well as those from other relevant writers, were then added to the collection. Although this archive with its dedicated building materialised only in the late 19th century, it was the first literary archive ever set up in Germany, strongly rooted in the emerging national feeling of the early 19th century — and at core drafted by none other than Goethe himself.[19]

Goethe had full authority over his papers as long as he was alive, just as he had full authority over his book production. This seems logical and legitimate: after all, it was his work. Who would be better entitled to make these kinds of choices over his production? Conversely, how can we be sure that a situation in which the author — in this case, an old man, and one of unequalled literary reputation — organised the archive would provide readers with a candid perspective on the work it presents? We cannot. There are plenty of reasons to believe that Goethe was keen on transmitting an ideal image of himself, an ideal he thought he owed to the nation, more than a reality.

Authority in the archiving process means projecting representations into the readers' expectations, and preparing to offer what one thinks it is that others would expect to find, trying to take into account the intentions they are likely to have. In other words, archivists, as representatives of an institution, aim to provide readers with what they think will interest them as regards the writer in question. In that, they stay true to archivist Hilary Jenkinson's position that evidence is key, and to that of Schellenberg, which states that the whole life cycle of documents has to be considered, meaning that they can have different functions and values at different

---

18   See section 2.1 for more on this aspect.
19   See *Goethe- und Schiller-Archiv* [50] on the history of this archive.

times. A handwritten letter, for instance, is first of interest to its recipient, but it may later turn out to be of interest to a wider audience.[20]

When the writer is the same person as the archivist, they have a unique overview of the evidence that is available, but at the same time, they might want readers to focus on some aspects of their work in particular, and be tempted to encourage a direction that suits them best. Whatever the collection, the person who has authority over the archiving process has to assess three fundamental issues: the source elements (indications given by the writer, for instance regarding relevance — in Jenkinson's words, the evidence); the target elements (expectations of the readers); and the institution's goals (specific missions of the institution hosting the concerned archive) — it is the latter two that define the scope of the document's secondary value according to Schellenberg's model. Statements of purposes for archives are key to understanding not only what they are keeping, but also why they are keeping it. There is not much other discursive space for archives to express the type of authoritative intervention that is at work in the processes of stocking, appraising, recording, and making available that they implement in shaping the character of their material.

The institutionalisation of archives over time has led to practices and to theoretical approaches in the archiving process that are strongly rooted in the materiality of the archive: what space it occupies and how; what role individuals and societies play in archiving as a process; and how archives are structured in relation to political orientations that impact social structures for decades.

The overview presented here provides a broad ouline of the way in which archives were institutionalised and how the definition of their missions evolved over time. In section 1.2, I will go into the archiving process of stocking, recording, and making available in more depth and from a more theoretical perspective. But, before delving into this topic, I will devote the next section to a general discussion of the elements of continuity between analog and digital archives.

---

20   See Cook, "What is Past is Prologue" [38], especially pp. 24-30 on the Jenkinson vs. Schellenberg debate.

1.1.2 Digital archiving

The emergence of IT-based technologies not only changed some of the technical infrastructure that hosting institutions can rely on: it also introduced a new meaning for the terms "archive" and "archiving" in the digital context. By "digital", I mean supported by a variety of devices able to perform computational tasks, such as computers, laptops, and tablets, connected to the internet, and in which content, including text, is represented by data.[21]

In the digital context, archiving means storing information that is encoded in such a manner that it can be read by a computer and retrieved at a later point. An archive, in the sense in which it is used in connection with IT technologies, provides a set of machine-readable information that makes it possible to retrace the evolution of a digital file. A digital file is the equivalent of what I called a document, or a manuscript, in section 1.1: it can be the letter you find in a drawer in your grandparents' home, or a random file you discover on a hard drive. In terms of its structure, a digital archive relies on the provision of different versions of the same reference document, making it possible to retrace the different steps of its completion up to the point when it is consulted in the present time. The process by which one keeps track of these different versions of the same document, called versioning, is key to digital archiving: it consists of storing information on a series of iterations of the file that reflect its evolution. Proper versioning also requires providing a hierarchy between the different versions (especially a temporal one: which version was there first, which one second, etc.) in order to facilitate navigation between these different versions. Each digital archive is a highly structured document, containing both the content of the file (data) and the information that makes it possible to retrace the evolution of the document's content (metadata).

Not that metadata is irrelevant for non-digital archives. All heritage institutions, including archives, museums, and libraries, have been working with metadata from the onset, and certainly already so in Ancient Athens or Egypt. Metadata are essential to any form of recording: what is being recorded by an archive is, primarily, metadata.

---

21   Data are organised in files, themselves structured in folders: in this area, the vocabulary is similar to that of the analog structure of document collections.

Metadata are "data about the data" — information about content, physical appearance, provenance, even in some cases the significance of a considered document or artefact. For the manuscript of a 19th-century letter, metadata will indicate that it is a letter; who the sender was; who the recipient was; when the letter was written (and, if known, when it was received); where it was sent from (and maybe where the recipient was staying when it was received); how many pages it has; on what paper it was written; whose hand has been at work on this manuscript (with what, ink or pencil, for example); which institution is currently in possession of the document (maybe also who has owned it before that); what kind of traces (stamps, foliation numbers, comments) were left on it in the course of archiving processes and by whom; whether the letter answers an earlier letter; whether there exists a response to this letter; whether there exist copies (maybe a printed version); and where those are to be found. These are the most basic metadata to be considered for a document like a manuscript of a letter. This information can be extended to a variety of more specific elements. And the answers provided to the questions asked — the metadata fields — can also be either very basic (just a name, a date, or a city), or elaborated upon in order to contribute additional context. Metadata, in and of itself, tells a story about a document, and this story can range from scant bits of information to a fully fleshed narrative. This is as true for digital files as well as for traditional archived material.[22]

Digital archiving relies strongly on metadata to identify different versions of the same document. Transposing the process of creating a digital archive for an analog document, taking again the example of the manuscript of a letter, the digital archive of a letter could for instance be structured along the following lines: first, include the initial text of the letter as it was drafted by the author when they sat down to write in the first place (version 1); then the version including the edits made by the author when they read what they had written and made some corrections to the letter before sending it (version 2); then the version deposited in a folder for preservation purposes, for instance, the box in which the recipient bundled all the letters from this specific correspondence partner, which is likely to include information on the position of the letter within

---

22   In *Managing Electronic Records*, Philip C. Bantin underlines the importance of metadata as a pivot between analog and digital in archival processes [20]; see especially p. 12.

the whole corpus, for example, its position in the chronological order of the correspondence (version 3); then the version generated when this correspondence folder was acquired by the archives and included in the structure of their holdings (version 4). In all of these versions, the content of the document (in this case, the wording of the text) varies very little. Also, most of the metadata describing the content of the document (sender, recipient, sending date, sending place, etc.) remains identical, although the information on the document, and the way it is part of a context, is susceptible to change, depending, for instance, on who acquires the document, how, and where it is preserved. This information can be recorded either by repeating all the information over again, or by identifying what varies from one iteration to the other and recording only this information. Here, just as with the analog context, there are choices to be made and informed decisions to be taken when shaping versioning mechanisms.

Versioning primarily involves the metadata related to the digital file: when the status of the file changes (when it is shifted from one folder to another, for instance), there is new information to be provided in terms of versioning. Digital methods have been created with the goal of recording only the elements that change from one version to another, and keeping the rest "as is" without having to record it anew. This allows one to identify differences between different versions swiftly. In order to gain yet more orientation, there exist techniques to structure the versions depending on the scale of the changes operated: master versions are the most important ones. This type of information provides readers with input on structure and hierarchy when they consult one specific version.

If we were to try to record every single version, every single change in document content or document status in every single digital document, this would lead to an information inflation of the same range as if we were to claim to archive "everything" in the analog world. There is not enough (virtual) room to keep track of every change that is made to every digital file, just as there is not enough physical room to keep every document ever written, including information on the context of its creation. But in terms of scale, things are rather different. Proper versioning, particularly in parsimonious computer languages, can be undertaken in such a way as not to take up much (virtual) space.

Here we can try a thought experiment. We can pretend to save an immense number of archived files on a variety of devices or cloud hosting, and assume that we would have enough server space to proceed with such an encompassing endeavour.

Even then, we would have trouble accessing information. Just as we might not be able to read some old handwritten scripts that have gone out of use, and need a Rosetta stone to help us translate them into languages we know, so computer-readable languages age and can become less and less readable over time. This is particularly true for proprietary binary formats developed by companies who make money from improving their products in such a way that they become more and more difficult to compute by standard machines, until one day, after yet another update, they become completely unreadable. Binary formats like those used in the office suite, for instance, are problematic because they embed raw text information and the formal features necessary to display it, making it impossible to access the raw text without the paraphemalia within which it is intertwined. This problem is well-identified, and there exist machine-readable languages that are of special interest because of their sustainability over time, as is, for instance, the Extensible Markup Language (XML) that has proven its stability over several decades. In its display of information, it separates semantic content from all that relates to formatting, but still makes the connection between them visible. More generally, open and free formats are conceived so as to be easier to adapt across technical evolutions, so that they remain readable.

Archiving strategies for digital files concern the infrastructure that supports them, but just as much as the computer language in which they are written. This directly concerns resources that need to be converted into a format that enables browser-based visualisation, such as HTML or formats that need to be convertible to HTML to be displayed. They rely on the development of technologies necessary for web visualisation and are susceptible to disappearance if one or the other of their features is not supported, for instance after a browser update.

Laments about online data ceasing to be accessible from one day to the next have become a topos of digitisation sceptics, often without any understanding of the clear distinction between the online display of the information and the source for this information, that is, its source code.

Let us consider the first type of online information: webpages that cannot be accessed after a while. The Internet Archive has proposed a new way of archiving webpages to address this problem. It implemented a tool called the Wayback Machine that makes it possible to go back in time to consult URLs in the way in which they were presented earlier.[23]. What is displayed by the Wayback Machine, though, is only the interface, that is, what is displayed on one specific URL at one specific point in time — like a screenshot from the past. It is not the original information (or source code) that is being archived, and it does not automatically reproduce its overall architecture including all the internal hyperlinks. The Wayback Machine does not archive the content of a database; it archives its presentation in a browser — it is not strictly speaking a digital archive since the source code is not what is archived, even if the HTML version is well preserved and can, to some extent, even be browsed. Although it has its limits, the Wayback Machine remains an endeavour of an unequalled scope. To fulfil its mission, it needs to keep browsing and harvesting the web for images of webpages whose content is likely to evolve over time — and that potentially means every single webpage, even if only because web design evolves and webpages follow this evolution. Of course, this is not done by hand, an automated process is at the core of this massive harvesting of web information. And it might well be that the precise snapshot of the precise URL you would like to consult has not been archived, and you are left with no result for your request. Nonetheless, the Wayback Machine remains a major asset when it comes to archiving internet-accessible content.

What it makes available for consultation, then, is only what is displayed, not the actual content of all of these webpages. This needs to be explained further, since it is essential in order to understand the structure of digitally available information in general. The key here is to understand the difference between, on the one hand, the information that machines display online, in general visually — their output, and, on the other, the information that they are provided with in order to display what they display — their input. Input and output provide different information and they do not rely on the same technologies to be accessible. Bridging the gap between them requires additional technologies to come into play

---

23   See `https://archive.org/web/`.

to process an input and transform it into an output. Yet other computer languages are used to process information that presents itself in a specific format or language, in order to make it readable, especially by humans.

One of the reasons why a language like XML (or LaTeX, which I am using to write this book) is so interesting is that, although they are used for writing source code (input), they remain readable both by humans and by machines. In these formats, text content coexists with information on how to display this content, all in a single file, although separated into two information levels (contrary to binary formats, which merge them, like a word document). It requires an additional intellectual effort to consider the different information layers presented in a file written in one of these languages, but it leaves room for interpretation when converting the information into a format intended for display. "What you see is what you get" text editors depend on display output. Moving away from that to start using languages, even as simple as Markdown[24] in which you can separate the content and how it is displayed, is certainly one of the most challenging aspects of digital information structure for non IT-specialists. It takes a while to understand that you can label a title line with a "title" tag, and decide later whether you want the title to be displayed centred, bold and red, or on the left, in small caps and black. You cannot see these output options in the basic text file as you do in a so-called WYSWYG text editor ("what you see is what you get", displaying from the onset the output layout), or, if this information is included, it is not intertwined with the text itself, but located in some other part of the digital document containing all the metadata pertaining to output display. Another advantage of these computer languages that describe the way the text presents itself is that they structure it. A digital file in XML, for instance, is organised as a tree with different branches containing information, all related to a common trunk. The hierarchies that structure the document follow the same tree-based logic, making it possible for a reader to easily gain orientation.

There is some complexity in digital archiving that goes beyond that of analog archiving. Taking the aforementioned aspects into account, what would an ideal digital archive look like? Ideally, a digital archive is presented in a stable language, saved on a sustainable server, and is so

---

24 See the Markdown Guide for more information: `https://www.markdownguide.org/`.

complete that someone wanting to consult any version of the file will be able to do so using only the information provided by the archive itself. It will contain at least the source code (a computer-readable file) and the information on how to transform this source code into human-readable information (output) such as, for example, an HTML file that can be displayed in a browser window. In that manner, even a reader that is not familiar with IT-based technologies or encoding languages can find all the information they need to access the information that is digitally archived.[25]

In that sense, "archive" has a somewhat different meaning when applied to the files stacked on a hard drive or to the old papers you found in a drawer in the family home. And yet, these are not two separate worlds, especially as digital formats are now also used to preserve information derived from the analog world (like those old papers from the drawer). Both techniques converge in the methods used for the digital archiving (and digital publishing) of analog sources. What is more, analog archiving and digital archiving have basic processes in common. Both rely on structured recording, and on updating information to make it understandable by readers. The ageing process is to some extent different: in the case of the analog archive, consultation can contribute largely to decay because it involves physical contact, while it does not modify the digital archive substantially. The digital archive loses its accessibility rather because of a lack of technical support or infrastructure for some formats. In both cases, however, one could generalise, saying that the cause for the lack of accessibility is not intrinsic to the archive itself, but to the fact that it is medium-dependent.

Digital archiving offers more food for thought than merely a polarised opposition to analog archiving: as we have just seen, there are more similarities than might be apparent at first glance. I would now like to turn to the notion of long-term archiving to consider how it can be implemented, taking into account not just the overall structure of digital archives, but also the readability of code. Long-term archiving is to some extent a contradiction in terms. What this expression actually means for the digital archive is "preserving data in their original format for about ten to twenty years". Ten years is certainly not a "long" time

---

25   For a more thorough presentation of the technical processes at work in digital archiving, see Ciaran B. Trace, *Beyond the Magic to the Mechanism* [100].

when applied to archives as we have known them historically, some of them being centuries old. But realistically speaking, in the context of IT technologies, ten years is a long time. We cannot be completely sure today which standards will apply then, which infrastructures will be available, which languages and formats will have been developed that are likely to make one specific type of files easier to read — or on the contrary impossible to execute. In the case of binary formats, it is almost certain that files written in a specific version of this format will not be readable with the version that is running ten years later since change is the basis of the business model they rely on. In other words, it is extremely risky to offer long-term archiving for longer than ten years, as you cannot be sure to be able to deliver on such a promise.[26] This has much to do with the format chosen for archiving, however, because the core text content, in a basic text format with no structure at all but just the sequence of signs, will, on the other hand, remain. Raw digital text is persistent.

Another difference, maybe on a more speculative level, concerns uniqueness. In the analog context, it is a basic difference between archiving and publishing that, while publishing by definition multiplies a text, archiving does not duplicate, but stores unique copies. The concept of archiving — preserving and recording a unique document for later consultation — is somewhat misleading in the case of digital information for which data has to be actualised, that is, overwritten, even perhaps transferred to another format, in order to still be consultable. Digital archiving calls for iterations or instantiations of archival material, while the uniqueness of the medium is essential to the definition of the analog archive. Taking again the example of the manuscript of a letter, there is only this one manuscript of this letter, and if we (or anyone else, even the authors themselves) copy it, the result will not be the manuscript of this letter, but a copy of it. It might preserve the wording better, for instance in case the paper is fragile and words become illegible for one reason or another, but it is only a copy. In the digital archive, the notion of an "original" as opposed to "copies" does not provide much orientation — every version is to some extent a copy, but also an original.

Readability of the digital archive by the computer does not mean that it can or should be generated without human intervention. On

---

26   These remarks apply to so-called hot data, that is destined to be opened and used. Cold data on the other hand (which is preserved, but not accessed and executed) is easy to store — but this does not encompass the scope of archiving at large.

the contrary, someone has to organise a digital archive just as someone has to place material in an archive box, store it on a shelf, and record it in a catalogue. In both cases, human effort is essential for defining metadata that will make retrieval of the archived information possible — in one case, boxes with numbers, in the other, metadata fields. But, in both cases, human intervention is key, even if some work steps can be automatised in one form or another. Again, here, languages at the interface between human-based reading and machine-based computing are a way of integrating these different dimensions. They represent a strong safeguard against a separation of computer processes and human approaches, a way to integrate large quantities of material and still have a sense of their quality. They are also a way of keeping archiving, at least to some extent, a human-based activity.

Let us consider a little longer this combination of human and computer activity in the archiving process specific to born-digital documents, created digitally at the outset, because they bring about yet other challenges. Transposing archiving processes of analog artefacts into the digital world can induce an intermediary work step that includes some human intervention, such as structuring the recording process. But the human activity in archiving born-digital productions is less straightforward. Ephemeral digital performances are of particular interest in that regard. They can be grasped statically (with a screenshot at one point in time), as by the Wayback Machine, but any dynamical presentation, such as rotating 3D model, or audio or video content is bound to elude this form of archiving. Today's authors are well aware of the type of power that can be exerted through the discontinuous availability of online material. Literature Nobel Prize laureate Elfriede Jelinek, for example, is known for randomly uploading literary content on her webpage — and erasing it randomly — making monitoring her webpage a tiresome activity for the research team dedicated to her work.[27] While Jelinek's course of action is in many ways similar to that of Goethe contemplating his personal archive, the seemingly random availability of information questions the author's role even more deeply. Subversive strategies like this one, inherited from 20th century forms of opposition to authority, can also be integrated in auctorial archival concepts that take specificities of the digital media into

---

27   Jelinek's webpage can be consulted at: `https://www.elfriedejelinek.com/`.

account. But whatever the digital staging of control and power on the author's side, if the text was online at some point, then it can be retrieved.

Born-digital archives bring other challenges too, especially for heritage institutions. Authors leaving behind hard drives instead of typewritten manuscripts, which in the course of the 20th century have taken over from the handwritten papers or notebooks I mentioned above, do not make archivists' lives easier. Hard drives do not take up as much space on the shelves as paper, but they too require specific temperature and humidity conditions to be well preserved. And they wear out. They also can very well stop working for no apparent reason and then all access is definitely lost. They need to undergo regular backup on other devices, not to mention actualisations in updated file formats, since popular text editors are proprietary and do not age well. They need to be updated to the next issue regularly, often at a certain cost.[28] In terms of curation, this can hardly be considered as an improvement compared to putting a stack of paper in a box on a shelf.

Questioning the quantity of information is all the more essential as digital archives are prone to never coming to an end, never being fully completed. There is no physical limitation to enriching a digital file, as opposed to a page that is, at one point or another, full. As a logical consequence, there is no need to consider that a digital resource is ever in its final and definite state. It can always be useful, compulsory even, to reopen and modify the file, version it and archive the new version.

One could certainly try to argue that the digital media does not require you to redraft the whole archiving process from scratch when dealing with material that was originally archived in an analog format. You may think that simply adapting this or that aspect (presenting a catalogue in a spreadsheet format for instance) suffices to achieve the shift into the digital.[29] When you consider what archiving means in general terms, the digitisation of archiving processes actually leads to major epistemological challenges, contradictions even. It involves changes in scale (the digital

---

28 Ageing concerns hard drives especially in the current context of global markets following a logic of planned obsolescence. As well as hardware, it also concerns software, i.e. the source code and the code that is necessary to execute it to provide access to file content. The Software Heritage project is dedicated to preserving software that is not in use any more. It curates source codes in order to provide long-term access to software content. See `https://www.softwareheritage.org/`.

29 See Glenn Dingwall, *Digital Preservation: From Possible to Practical* [44].

space being seemingly much more vast than the physical world, and yet, in another way, finite); the move to a duplication of the material (the variety of digital versions when a unique manuscript used to be the rule); the emergence of archival practices in a wide array of areas (archiving all kinds of material); and new approaches to text as a representational form (specifically in the archiving process itself, with the separation between source code and visual output).[30]

Every archive is a human-made construction of textuality at large. Digital archives are such a construction— perhaps even more so than any other — because of the balance they strive to maintain between contraction and expansion, preservation and accessibility. Broadly speaking, an archive is composed of different text layers, including data and metadata. We need to know how they are situated in their specific space and time: that is what gives readers a chance to position themselves in their relationship to these documents or files, be they old papers in a drawer or a digital file on a hard drive. We need to be able to state where we speak from when we speak about one specific piece of archive.

1.2 From writing to reading to archiving: becoming text

In the previous sections, I have considered archiving processes and their challenges in the analog context and in the digital context, showing how these processes are in some ways similar, but in others divergent. A key element on which to ground this comparison is the textuality that both processes rely on: the letters found in a drawer, the file on a hard drive, even to some extent the source code, are forms of textuality that have to be preserved to be made accessible.

The following section explores what text is made of more fully. As a first step, I look into the relationship between media character and temporality in text genesis. Then I put forward a step by step reconstruction of how text becomes a text, and through which process. Finally, I will consider the archival temporality specific to the preservation of textual traces.

These analyses aim at providing a theoretical framework for an approach of text that encompasses archival functions and other forms of

---

30   The principles I have presented in this section as an argumentative narrative can be found in a synthetic form under the title "Ending Principles for Digital Longevity" at `https://endings.uvic.ca/principles.html`.

public availability, and positions preservation challenges in this overall framework.

1.2.1 Temporality and text as media: the power of genesis

I have mentioned examples of authors who have attempted to monitor how their work would be archived: Goethe by gathering all kinds of papers at the end of his life and having them arranged by dedicated secretaries; Jelinek by using her webpage to create ephemeral archives of her writings. While these practices follow decisions taken by individuals, they also reflect the archiving habits of specific periods more generally, which are themselves strongly connected to the political context of each period. The contrast, even if only between the 19th and the 20th century, is striking. Manuscripts of the 19th century are often thoroughly doctored or even rewritten by complete strangers.[31] In the 20th century, it has become common practice to preserve, order and transcribe manuscripts by the letter, avoiding interventions. French critic and editor Jacques Petit considers authors who personally see to archiving their manuscripts as particularly affective for their readers.[32] He speaks of a dizziness ("vertige") seizing whoever might try to take a stance on any part of the archival universe of an author without taking into consideration all that the author chose to preserve. The underlying assumption is that, if an author preserves all of their manuscripts, readers should go through all of them if they want to form an opinion on the author's work or even dare to make an interpretation of part of it. The commentator would have to take into account everything that had been written (and archived) by the author, paying attention to the wealth and variety of textual material available, and could only legitimately attempt an interpretation after acquiring a knowledge of the author's entire production. Petit suggests that the preservation strategy chosen by these authors is a way for them to pressure their readers (maybe more so well-read critics or literary

---

31 See for instance Sophia Zeil's and my analysis of interventions in the context of the *Tieck and Solger* corpus [19], or Anna Busch and Johannes Görbert's of Chamissos *Weltreise* [34].

32 Petit writes about early 20th century authors who archived almost all their work: "Et lorsqu'un écrivain comme Mauriac, Claudel, la plupart des contemporains, a conservé presque tous ses manuscrits, on se sent pris de vertige en constatant que tout ce qu'on en dit sans avoir *tout* lu, est *un peu* faux.". See "Pour une typologie des textes littéraires" in Hay, *Avant-texte* [66], p. 194.

scholars than "simple" readership though) and challenge their ability to embrace the whole of even a sole work. But does keeping all manuscripts, all drafts, all different states in the production of a text mean that every one of these work steps has to be taken into account at all times? Does it mean that it is necessary to know all of what is left in order to be able to address, consult, comment on a single text of an author? What do these diverse fragments mean in the overall puzzle of an œuvre, and what is their specific significance?

It is not technically possible to take into account every component of a textual ensemble every single time we read a text or every time we are confronted with a loose manuscript. What we interpret is generally one specific state of a text, one moment in its editorial history — the first edition of novel, for instance, or the primary draft of a manuscript. These are not identical with the second edition (the layout may have changed, typographical errors will have been corrected), or the following draft, which will probably have fewer edits. Hence the question: how can we address the fact that a text does not emerge suddenly like Athena, springing fully armed from Zeus' head, but has a history of becoming a text? How can we define it in such a way as to be able to situate our approach for each specific stage of the text, at each specific time in the process of its creation?

Depending on the consultation context, the position occupied by a specific version of a text can change. The materiality of the manuscript evolves — paper ageing, ink decomposing paper, pencil naturally erasing. The act of consulting a text, most of all a manuscript, has an inherently transitory quality: each consultation reveals a different iteration of the text, as the media it is inscribed in keeps evolving over time. Considering this changing situation, it seems necessary to address systematically what temporality, in relationship to its media, defines each moment in what we consider a text. Or does this inevitably lead to making a text eventually impossible to grasp? That is: does asking myself the question of what stage of the text I have in front of me make my reading easier and more legitimate; or, on the contrary, does it make reading it unnecessarily complex, even impossible to proceed with? What is it I hold when a manuscript passes through my hands? How can I know whether it is of major importance or virtually irrelevant?

The more one tries to grasp what constitutes a text as a part of a more
general context, the more it slips through one's fingers and seems to
lose its substance. Here, I will take the example of a literary œuvre for
simplicity's sake, but will elaborate on the difference between text, book
and œuvre in 2.1. The French school of the *critique génétique* constructed
a differentiation between *avant-texte, texte* and *après-texte* (before-text, text
and after-text).[33] Although this three-step structure provides helpful
orientation, it lacks consequence in at least two points. First, it imposes
a clear unidirectional linearity: the three steps follow one another and
cannot be shuffled into a different order than that of before, during and
after the composition of a fixed, core version of a text. And second, it
suggests that the *avant-texte* could not exist without that which comes
after it. The denomination *avant-texte, texte* and *après-texte* includes a
pre-determined teleological bias as it revolves around the idea of one core
text, chronologically situated in the middle. In these terms, one and only
one version of the text gives meaning to the other versions. In practice,
this ideal framework is not that simple to apply. There are often various
copies or versions of a manuscript or a print, and one cannot always easily
settle which one is authoritative, which is "the text".

Defining text is extremely easy and extremely complicated at the same
time. French tradition of the late 20th century provides a few definitions
that hardly resist media variability as we know it today, but still offer
food for thought. Philosopher Paul Ricœur states: "Appelons texte tout
discours fixé par l'écriture"[34], calling text any discourse fixed by writing
— the focus is on the writing process, insofar as it is achieved. Derrida too
focuses on writing, but he emphasises how it is rooted in origin rather than

---

33   In the eponymous seminal volume [66], the article by Jean Bellemin-Noël, who first
     proposes a definition of the "before-text" that struggles to emancipate from the notion
     of a "draft" that would simply be a prior, less achieved version (p. 162), but he
     then proceeds more convincingly: "Il est indispensable de substituer les métaphores
     spatiales aux images tirées d'un registre temporel, sous peine de réintroduire une
     téléologie, alors que la rédaction n'a pu être qu'une production toujours surplombée
     par l'incertitude et l'aléatoire. Le paradoxe à ne jamais oublier, le voici : ce qui a été écrit
     avant et qui n'avait a priori pas d'*après*, nous ne le connaissons qu'*après*, avec la tentation
     d'en faire un *avant* au sens de préalable, de cause, d'origine." (p. 163). Bellemin calls
     for moving from a spatial to a temporal register and sees it as an essential paradox
     that what was written "before" did not have an "after" when it was written, adding
     uncertainty and randomness to the process. In recent work by the *critique génétique*,
     even more complex concepts and vocabulary were developed, which can be consulted
     in the *Dictionnaire de critique génétique*: `http://www.item.ens.fr/dictionnaire/`.
34   Ricœur, *Du texte à l'action* [91], p. 137.

achievement: "'Qu'est-ce que l'écriture' veut dire 'où et quand commence l'écriture'"[35] — "what is writing" means "where and when does writing begin".

This remains general and cannot easily be transformed into a practical way of reading or interpreting texts. Anglo-American textual studies provide a more pragmatic grasp on socio-historical factors without giving up on a theoretical framework. American textual scholar David Greetham writes:

> [The text] is, on the one hand, a place of fixed, determinable, concrete signs, a material artefact, and yet, on the other, an ineffable location of immaterial concepts, not dependent at all on performance transmission. It is, on the one hand, a weighty authority with direct access to originary meaning and, on the other, a slowly accumulating, socially derived series of meanings, each at war with the other for prominence and acceptance. It is a place inhabited only by a sole, creative author who unwillingly releases control to social transmission, and it is also a place constructed wholly out of social negotiations over transmission and reception.[36]

Greetham argues along the lines of authority, control and concurrence. The metaphor of the battlefield is not the most reassuring one, but there truly are troop movements in the text: the author's, those of the reader(s), the pre-existing meaning of the words that are used, and the materiality of the text itself. Greetham describes them in a fight for attention, for authority within the text. There are different, sometimes opposing forces at work. Their plasticity is made all the more mouldable as they take account of the media options that are available. But these are not the sole driving forces at play. Greetham also emphasises the effort required from the author in order to enable transmission. Text happens between participants, and the "negotiations" take place both on a social and a material level.

The definitions I have cited all agree on the idea that text is in movement, both on the space of the page (the discourse "is being fixed", following Ricoeur; it "begins somewhere" following Derrida; or "is a

---

35   Derrida, *De la grammatologie* [41], p. 43.
36   Greetham, *Theories of the Text* [65], p. 63.

place", in Greetham's words) and in the temporality of it being written and read (Derrida also asks "when" it starts; Greetham suggests a "slow accumulation").

One way to focus these approaches would be to say that there is something like a geography and a history of a text, both anchored in its materiality and how it evolves in terms of media.

We can turn first to what I want to call its geography, or topography. It pertains to the way in which the signs are disposed on the page, and designates the topography of the chosen medium. Prosaically, it describes how darker and lighter bits — such as black ink on a white ground — alternate to shape letters and other glyphs on a page. It is not the letters that are written which produce meaning, but the way in which the different elements contrast with one another on a given space: it is the composition that matters. The topography of a text varies very little if at all when you compare different copies of the same edition; it varies a little more when you compare different editions of the same book; it varies remarkably when the two versions compared to one another are based on different media (for instance, a manuscript on the one hand and a printed copy on the other). Poorly executed print-on-demand is a good example of the fact that reproducing pages identically does not necessarily convey the same reading impression: bad print, imprecise reproduction, and lesser readability are more the rule than the exception in these practices. What this example also shows is that the geography or topography of a text is closely connected to the type of writing or printing equipment that is available at the time in which a text is either written or reproduced: ink type, paper type, printing equipment, and fonts do not vary randomly, they reflect the conditions in which the text is being produced.

For instance, Gothic print is strongly rooted in the development of print in the German-speaking area. Nowadays, these fonts are generally associated in the collective imagination with nationalist claims, as if they would programmatically incarnate the distinction from Latin print, and 'Latin' countries. In extreme cases, Gothic print can even be interpreted as an affirmation of the superiority of Germanic culture. Yet their cultural meaning used to be quite different. In the 18th-century German-speaking world, Latin script was considered a scholarly print, and more difficult to read for readers from the German-speaking areas. Gothic print was used for printing leisure reading and deemed more likely to foster pleasure in the reading process. This was the dominant point of view in the period when the publication market expanded dramatically, with the increase in leisure reading in the late 18th to early 19th century. A few German printers such as Unger, together with French printer Didot, engaged in creating a new set of printing letters (so-called types) that would combine characteristics of Latin and of Gothic print, but their attempt failed.[37] It would certainly have required at least a strong political will, if not an economical investment, to consolidate a technological effort that would eventually have facilitated the circulation of German prints on the European book market for many decades. Instead, by the middle of the 20th century, the cultural reference was not any more that of scholarly print vs. leisure print, but that of Latin vs. German print, with a strong political connotation. The materiality of text is itself political.

Choices in the elements that compose the print such as type or ink are related to what I have identified as the geography or topography of text. In addition to this dimension of space, each text has a dimension of time (what I would call its inherent history): text evolves over time. When textual geneticists state the distinction between *avant-texte*, *texte* and *après-texte*, they also suggest that one of the speculative difficulties is to consider this temporality in itself, to take into account the fact that the evolution of text is not necessarily organised along the lines of a straight time arrow, but is susceptible to leaps of time — in one direction or another —, affecting text and its materiality. The different steps in the dissemination of text are one possible matrix here (from the draft to the publication), but they should not be considered as an exclusive linear prescriptive power defining how text evolves, comes to being and is to be interpreted at all times.[38] Considering the evolution of text within time and yet not necessarily in a linear way is all the more difficult because we are used to reading in a linear order. It is rather speculative to stop believing that things evolve along a unidirectional time axis, but that is the idea proposed here: to try to reconstruct the inherent history of each text in a temporal framework that is not always moving in one direction from a beginning to an end in a direct line.

---

37   See Lehmstedt, *Ich bin nun vollends* [76].
38   See Vasak, "Analyse de système et textologie", in: Hay, *Avant-texte* [66], p. 199.

What I call the inherent history of a text commences with the act of writing in itself. Even this initial writing process can have a different significance depending on the author. For some writers, even the first draft presents some degree of achievement: they are *Kopfarbeiter*, people who draft mentally before starting to put anything down on paper. For other writers, it is the writing itself that initiates the process of drafting and redrafting until the meaning of the text emerges. The act of writing, as the first moment in the history of a text, informs the text both in terms of its content and its shape. A short poem will take little space on the page and be divided into stanzas. A prose narration will occupy the page space more completely. A letter will have a dateline, a salute, and the centred name of an addressee. While these elements can change at a later point (just as content can), they are drafted through this first writing act.

It is tempting to consider this stage of initial writing from a teleological perspective, that is, to interpret all the signs that are visible on the paper in such a way that they tend to become what the text will look like in its final stage. Reading into what one can see in a first draft, one could be tempted to interpret it along these lines of: here, the author started with this word, but then they preferred this other one because it addressed their intention more clearly, as the final version shows. Much can be projected in the interpretation process. Ricoeur objects to the risks of too strong a psychological interpretation.[39] The writing process itself, says Ricoeur, gives the text autonomy, disconnects it from its author's intention. The conclusion he draws from this observation is that psychological, as well as sociological creation conditions are insufficient to account for textuality. Textuality transcends its own creation and gives way to myriad readings, themselves anchored in different socio-cultural contexts. This argument is rooted in a reflection on the work of art in general. At its core, it calls for dissociating text from speculations on the author's intention from the onset of the writing process, and that is exactly what matters

---

39   Ricoeur, *Du texte à l'action* [91], p. 11: "l'écriture rend le texte autonome à l'égard de l'intention de l'auteur. Ce que le texte signifie ne coïncide plus avec ce que l'auteur a voulu dire. Signification verbale, c'est-à-dire textuelle, et signification mentale, c'est-à-dire psychologique, ont désormais des destins différents. [...] ce qui est vrai des conditions psychologiques l'est aussi des conditions sociologiques de la production du texte; il est essentiel à une œuvre littéraire, à une œuvre d'art en général, qu'elle transcende ses propres conditions psychosociologiques de production et qu'elle ouvre ainsi à une suite illimitée de lectures, elles-mêmes situées dans des contextes socio-culturels différents."

here: to consider a text as not strictly or exclusively correlated to its writer at the moment of writing.

There would be much to say about the initial moment in the inherent history of a text. The moment when writing begins is often interpreted in terms of the continuity and/or rupture elements that can be spotted on the media device —I will speak of a page for simplicity's sake. Does a variation in ink colour mean that the writer changed their mind? Does it mean they took a break at this point in the writing process — but if so, why? It is rather difficult, when considering the initial stage of writing, to actually dissociate text from author, materiality from individual. The writer is, after all, a human being, who sometimes gets disturbed by someone knocking at the door, just as much as they are a poet who needs to reconsider a verse for reasons of rhythmic balance. Who are we as readers to decide whether the author had a deep reason to pause or just had their cat knock over the inkwell? The question then is how we integrate the space outside the page, where things happen that we do not know much about, and what do we make of the signs we see on the page, or what we can see of them. One way of avoiding an over-interpretation of such text stages that would be too strongly led by authorial considerations is to primarily take the hand(s) that wrote into account, and not the person. For a draft, this approach facilitates interpretation.

One might object that not all writing processes formally require a hand and a paper manuscript. The hand considered here is not to be taken literally, but rather as a symbol for becoming written. At this point in the process, the text is not really yet a text, in the sense that it has not yet been read by someone other than the writer: the hand needs another person's eyes to realise the textual potential. This is, in fact, the next step in what I call the history of the text. The media history of a text does not fully take shape at the moment of writing, but in the next step, the moment of transmitting, of giving it to someone else to read. The transmission of a text can happen in a variety of contexts: in a private circle or to a public audience, in the form of manuscript or print. What these contexts have in common is the shift to a reader that is not the author — eyes that do not belong to the hand.

Writing for other eyes often means re-writing for the purpose of communication: what was a draft becomes a cleaner version, with fewer edits and more space dedicated to the final shape of the text. Some authors write their primary drafts using abbreviations: in the version for another reader, these will have to be expanded so that complete words are legible to those who are not familiar with the abbreviations. This first "clean" version of the text can be carried out by the same hand as the initial draft, but it can also be another hand's doing. Authors who can afford to dictate their initial draft to a secretary are less common than those who commission a third party to realise copies of their initial drafts. From a certain age, Goethe dictated almost everything he wrote — and yet one cannot say that he did not write.

The choice of the hand that will copy the draft has varied a lot over time. From the late 18th century and the development of editorial businesses as family endeavours, it was often a wife or a daughter who was in charge of copying, the activity that was considered not to entail any creative dimension. Wealthier writers could employ someone, a student for instance, or a professional copyist, to that end. It can be argued that this writing step, that of copying (in the literal sense of procuring a copy), is more creative than might be assumed: it is undoubtedly a material change that is likely to induce variations, shifts in the text. On the formal side, the disposition of the text on the page will most certainly change. In terms of content, it is prone to introduce errors (in the case of an erroneous reading of the draft, for instance, or a mistaken transcription by the new hand), or even intended corrections (if the copyist adds what they consider is a missing word in the text, or deletes one that seems superfluous). The *New Bibliography* differentiates, from this stage on, between *accidentals* and *substantials* in order to characterise the types of variations that can be found in a text. Some concern mistakes, typos, or other minimal alterations, while others involve more considerable shifts in the fabric of the text.[40]

Ideally, it should be uncomplicated to draw the line between modifications that qualify as *accidentals* and those that are *substantials*. But a great deal of textual phenomena operates in the grey zone in which the distinction between the status of each of these is not unequivocal, especially when it comes to stylistic alterations. Novels from the 19th century first published, often in form of episodes in journals, when their authors were young and penniless, could undergo a major rewriting process for

---

40   The importance of the New Bibliography is pointed out by Greetham in *Textual scholarship* [63].

later editions. This not so uncommon situation gives a sense of the type
of challenges faced when interpreting these different versions of a text
that has the exact same title. Greetham writes:

> [...] several major Victorian authors took advantage of reprints
> and new editions to rewrite the accidentals, often to undo
> the house-styling that had been foisted on their texts when
> they had been young struggling artists having to capitulate
> to publisher's styling demands. Dooley's argument is that
> the balance between intention and expectation has to be ad-
> judicated for each text (and that includes each edition), for
> the equilibrium shifts as the relative power of author and
> publisher changes during an author's career.[41]

At this point of visibility or public dissemination of the text, the relationships between writer and publisher come into play. I will explore them
more in detail in 2.1.1: encapsulating their significance is not as simple
as it may seem. While the publisher is not the author, they can have a
major impact not only on choices in layout, design, fonts, bindings, etc.,
but also on the wording. Looking at it, not from the point of view of the
actors involved, but from the point of view of the text and its inherent
history, what is at play in the intervention of a publisher is the relationship
between primary intention and authorship. Who can be considered as
the author, of, say, a printed version of a text in which the publication
process has involved many variations in the wording, when compared to
the manuscript that was sent by the writer in the first place?

These tensions are sometimes displayed very prominently in printed
products, particularly books, as is the case where one finds *corrigenda*.
They generally consist of a list of items that were printed incorrectly, such
as word omissions or mistaken names that are likely to change the reading
one can make of the text. They are presented together with the correction
wanted by the author and the number of the page where the error is
located. When there are *corrigenda*, they are often displayed beforehand,
at the front of the book. The pre-eminence of this position, at the forefront
of the publication, can be considered to reflect the authority the author
has over the publisher: not only does the author re-establish the text as

---

41   Greetham, *Theories of the Text* [65], p. 194

it should be read, but they do so in the first place, sometimes before the title page itself. This is more than anecdotal. It shows that even in the printed version of the text, there is room for variations in disposition — for making text happen in one way rather than another.

Having distinguished a series of steps in which text is being modelled in such a way, first by a hand, then by alien eyes, and further a copying hand or a printing device, all contributing to making it become text, one last step is still missing. What happens when the text is being read, not simply by friends, family and publisher, but at large? This is the intersection between text genesis and text reception. Text genesis is defined as that which concerns the creation of text. Reception analyses how the text is being read and commented on by third parties. In general, the genesis of a text is interwoven with the reception of other texts: writers have read books before they themselves start writing. Any form of writing is always an intertext of some sort, as it is created in a context where other texts serve as a background that is more or less conscious, and more or less formalised as such, but that always interferes and contributes to co-create the new text in some way. The genesis of any text lies in the reception of other texts, and every text is always an intertext from the moment when it is read by someone. This intersection between writing and being written, reading and being read, generates a never-ending revolution in the circulation of texts, a movement that operates at the core of the history of any text.

This process affects all stages of a text and its publications, including editing, as Greetham concludes, and to some extent, as we will see, includes archiving as well:

> The business of editing, just like the business of literary criticism and the business of writing about textual and literary criticism, is relentlessly intertextual. For although there has always been the temptation for the editor to lay a deadening hand upon the text and fix it forever, such temptations will not withstand the forces of history or the enquiring intellect: these "fixed" texts will always have to be "done" again.[42]

Being aware that text is not fixed is central. One can usually spot movements or shifts in the writing process or in the transmission, that can be

---

42   Greetham, *Theories of the Text* [65], p. 461.

associated with writing hands at work. This helps to avoid authorial over-load in the interpretation and prevents readers from projecting intentions into stages of text constitution. But in the end, one should consider that any interpretative gesture can always be questioned by the emergence of new texts, new intertexts.

The manuscript is an essential object around which the history of the text can be spun, and with it, its existence in the world. It is the bearer of information about the text that archives can record and preserve. Depending on the archived stage, it can reflect the editorial process that transforms a text into a publication to a greater or lesser extent. This complexity is reflected in archival systematics, but it also leaves room for interpretation on the archivist's side. What copy is the authorial one? What hand is that of the author? These are the questions archivists too have to answer, and that serve as a basis for our access to text.

In this section, I have examined text genesis, the process by which text becomes text, emphasising on the one hand the plasticity of text (always liable to change), on the other hand how strongly rooted it is in its media form. In the following section, I will examine this from the perspective of the archiving process proper.

1.2.2 Traces of the past in the present

We can now turn again to the old papers you found in the drawer of the family home, the postcard collection you purchased at the flea market, or even the hard drive a writer has deposited in literary archives, and examine them in the light of the text genesis process. What you find can be a draft; or it may be a copy; or it may be a printed version — it can be all these layers and versions, comprehensively bundled in a folder so that the reader will be able to follow the textual evolution. It may be a fragment of something one cannot even identify: a note? the draft of a letter? part of a diary? Has it been read before, or am I the first one to lay eyes on it? All these questions require you to remember that you are dealing with more than a text, but with what has become a document in the sense that it has gained recognition as a structured part of heritage.[43] This status as a document primarily concerns the epistemological value,

---

43 A group of French scholars has published an encompassing theory of documents under the collective name of Roger Pédauque, see "Document et modernités" [87]. Ricœur considers the relationship between document and trace within the framework of history

the knowledge it holds. In terms of its materiality, it consists of what are considered traces of the past that have reached a reader, traces that are anchored in both past and present. The edges of the page are certainly not the same as when the text was initially written, and the ink is not the exact same colour; what has reached the present is only partially identical to what it was in the past. What is more, considering texts as documents in the sense that they are traces of something past assumes that not everything that was produced in the past has reached us today. Archives do not actually contain all that was ever written, but solely a fragment of it, traces of time past. How does a document-based approach relate to an understanding of fragmentary traces?[44]

This is a challenge to anyone interested in the past. Trying to disentangle its complexity, we can start by asking ourselves: What are these traces of? What do they represent in terms of our knowledge of the past and of ourselves? Ricœur articulates this crux in *Time and Narrative*.[45] The paradox he points to there is that the traces we can see bear witness to an ageing process of which we can only see the result, and not the process itself — and even less so the original state of things. The trace, considered as the physical evidence on which historical research is based, lacks existentiality: it is neither what it has been by the time it was created and was not a document of something past yet nor is it a real thing of our time. It is constructed as a thing of the past while it is, in fact, a projection of the present on the past. It is also what Derrida points to when he writes: "la trace *n'est rien*, elle n'est pas un étant" — "The trace *is nothing*, it is not a being"[46]. The trace is a symbol of something that is

---

as a documentary science: "L'histoire en tant que recherche s'arrête au document comme chose donnée, même lorsqu'elle élève au rang de document des traces du passé qui n'étaient pas destinées à étayer un récit historique. L'invention documentaire est donc encore une question d'épistémologie. Ce qui ne l'est plus, c'est la question de savoir ce que signifie la visée par laquelle, en inventant des documents [...], l'histoire a conscience de se rapporter à des événements "réellement" arrivés. C'est dans cette conscience que le document devient trace, c'est-à-dire [...] à la fois un reste et un signe de ce qui fut et n'est plus.", see *Temps et récit III* [90], p. 13. In his view, history invents documents based on traces of the past, but should do so knowingly. The *SAA glossary* focuses its definition of document on its medium more than on its connection to the past: https://dictionary.archivists.org/entry/document.html.

44 See Baillot, *Reconstruire ce qui manque* [10].

45 Ricœur, *Temps et récit III* [90], pp. 217–218: "C'est bien là le nœud du paradoxe. D'une part, la trace est visible ici et maintenant, comme vestige, comme marque. D'autre part, il y a trace parce que auparavant un homme, un animal est passé par là; une chose a agi. [...] Où est alors le paradoxe? En ceci que le passage n'est plus, mais que la trace demeure [...]."

46 Derrida, *De la Grammatologie* [41], p. 110.

gone, a *symbolon* in the etymological sense of the word, a token made of
one bit that is left behind where it originated in the past, and another one
sent to the future world bearing the remembrance of the missing bit. This
symbolon is only whole when both parts are physically united to form the
original item. But the first part is lost for ever, swallowed by time, and all
we have is the second one, always painfully reminding us that it is not
whole, but a fragment. From this perspective, the awareness of what is
missing, or at least of the fact that something is missing, makes it possible
to actually structure knowledge.

Derrida brings the paradox to the limit of knowledgeability.[47] The
trace does not only indicate the disappearance of what was at the origin,
it signals that it was never itself at the origin — it was something else. It
would, according to Derrida, require a completely different concept, that
of an archi-trace, to actually embrace the trace that is at the origin. In fact,
it is impossible to consider what the trace originally was, or to unite the
notions of trace and of origin. And if the trace has no clear origin, then
what is it a trace of? From this perspective, all our traces from the past
are only ever but constructs of something that is not original, and whose
relevance always needs to be reassessed.

And yet, for reasons sometimes unclear, something was left behind of
time past following selection mechanisms that often lack transparency,
and this trace remains the main artefact on which historical perspectives
on text rely. Due to its loose temporality, to the uncertainty of its origin
(where did it come from? what did it originally look like? what did it
mean?), it remains to some extent difficult to grasp. We do not ever really
know what it was the trace of. It is the trace of something that is not any
more, and we cannot know whether what we think it emanated from has
ever been, nor what it really was. All we know for certain, at a speculative
level, is that the trace is not that which it is the trace of, and that it lacks
its origin when it reaches us.

We do not know where the trace comes from, yet at the very same time,
it is the well from which we draw. As if this was not embarrassing enough

---

47   In *De la Grammatologie*, he writes: "La trace n'est pas seulement la disparition de
       l'origine, elle veut dire ici [...] que l'origine n'a même pas disparu, qu'elle n'a jamais
       été constituée qu'en retour par une non-origine, la trace, qui devient ainsi origine
       de l'origine. Dès lors, pour arracher le concept de trace au schéma classique qui la
       ferait dériver d'une présence ou d'une non-trace originaire et qui en ferait une marque
       empirique, il faut bien parler de trace originaire ou d'archi-trace"; see Derrida, *De la
       Grammatologie* [41], p. 59.

from an epistemological point of view, the question of its preservation is also full of contradictions. Keeping everything is impossible. Choices and selections are being made. According to Ricoeur, this selection is natural, or at least congenial to the order of things. It suffices to do nothing for things to become ruins.[48] Not taking action is leaving it to time to do the work of letting things decay — and believing it is time itself that brings about destruction.

But time is not the only factor. Other mechanisms are at work, connected to recording and archiving processes, as was mentioned in section 1.1.1. Archiving means selecting, be it unwittingly, if physical decay or loss could not be avoided, or on purpose, in order to keep some documents rather than others — family papers only if they are 150 years old, novel manuscripts only if they are by famous authors. These are deliberate choices connected to the mission of purpose of heritage institutions, a principle officially embraced by the profession at least since Schellenberg's stance on appraisal.[49]

Whatever the mission or purposes, there is common ground: the fact is that neither institutions nor individuals are in a position to keep all traces from the past. Some artefacts, some pages, will inevitably have to disappear. There is not enough room to keep everything, and not enough time to record everything so that it can remain consultable.

But at the same time, keeping order in the traces of the past to archive them, recording them, means proceeding to actions that will both preserve and destroy content. There always lingers above archival documents the threat of destruction; this, too, is part of what defines them.

The threat comes from natural (uncontrollable, irrational) and human (deliberate, motivated, even rationalised) destruction, both at work during the archiving process. French archivist Arlette Farge describes the frailty of archival documents almost poetically, each a "piètre pièce

---

48   See Ricœur, *Temps et récit III* [90], p. 34: "En un sens, il n'y a là rien de mystérieux; il faut en effet faire quelque chose pour que les choses adviennent et progressent; il suffit de ne rien faire pour que les choses tombent en ruine, nous attribuons alors volontiers la destruction au temps lui-même."

49   In *The Appraisal of Modern Public Records*, Schellenberg states this in unambiguous terms: "A reduction in the quantity of such public records is essential to both the government and the scholar.[...] Scholars cannot find their way through the huge quantities of modern public records. The records must be reduced in quantity to make them useful for scholarly research." [94], p. 237.

[...] à conviction, toute en dentelles", a piece of lace.[50] Similarly, German philosopher Knut Ebeling characterises archives as a "Friedhof der Schrift",[51] a cemetery of writings, and goes one step further in the register of destruction, up to institutionalised rot and decay. Derrida too comments on the fundamentally destructive power of archiving: "L'archive travaille toujours et *a priori* contre elle-même"[52]: repetition, inscribed by recording and preserving in order to consult archived documents, is the process through which the archive works "always and *a priori* against itself". Each consultation of an archival document contributes to its decay. This means that it is not technically possible to consult the exact same document twice: between the two moments of consultation, the document will have evolved, and will not be exactly the same any more. (One could also say as much of the person who consults it too: one cannot encounter a manuscript twice in the exact same state of mind and body.) Consultation, the only action that can contribute to actualising the potential of preserved archival traces, accelerates their degradation. It induces an uneven time trial in which the intelligence generated by consultation has to keep up with, maybe even overtake, the degradation process accelerated by consultation, which itself contributes to loss of information.

This perspective also helps to shed to some light on the contribution of digitisation processes in the archiving process of physical artefacts. Scanning a document or an artefact that is degrading at a dangerous rate — at a rate deemed dangerous in the temporality of cultural appraisal and research processes — makes it possible to record the evolution of this degradation in two dimensions and with a pre-defined image precision. On the one hand, digitisation thus makes it possible to keep track of the history of archival traces. Thanks to it, we have the opportunity not to lose completely traces that are decaying before our eyes. The physical artefact becomes more valuable in the sense that an in-depth interpretation, taking into account its evolution, remains possible. The material life of the

---

50   Farge, *Le goût de l'archive* [48], p. 73: here with the words "meagre evidence" and "lace-like". Farge also uses the image of a puzzle whose pieces are scattered and can only be put together imperfectly in the form of obscure events: "Si l'archive sert effectivement d'observatoire social, ce n'est qu'à travers l'éparpillement des renseignements éclatés, le puzzle imparfaitement reconstitué d'événements obscurs." [48], p. 114.
51   Ebeling, *Archivologie* [45], p. 12.
52   Derrida, *Mal d'archive* [42], pp. 26-27.

document can, to some extent, be accounted for, visually at least. On the other hand, however, digitising a source document can generate the illusion that the image and what it is an image of are identical. The original document loses symbolic value, the scan gains a validity that may even be considered absolute: you could argue that the scan is better than the original since you can zoom in and see details the naked eye cannot grasp. But in this the temporality of the archiving process gets lost. Because the fact is that very few people engage in an elaborate consideration of the date and conditions of digitisation when they work with a scan. Technically, the scanning date and conditions should be added to the date of the source document to really define the temporality that is the one of its material textual traces when it is consulted as a scan on a computer monitor or on a smartphone display.

Not only the consultation of documents is a cause for decay: when paper is exposed to air, light, or heat, its degradation accelerates. And even if conservation operates with all thinkable and realisable precautions — optimising humidity and temperature, keeping sheets in protective boxes, finely separating them from one another, away from light — it too becomes part of the degradation process. Conservation does not stop time, it only slows down some of its effects. Acid inks will still corrode paper, for instance. This is part of the natural course of things — "time" in the words of Ricoeur — but also of the use of degradable cultural techniques and artefacts.

In the archiving process then, an archive deploys mechanisms of conservation that include a dimension of destruction. One of the consequences of this is that one cannot always dissociate conservation from destruction. Not all the traces we would need to reconstruct what we want to know of the past are available. We do not even have enough traces to know for sure what the blind spots are, to measure the extent of what has been lost over time. Derrida addresses this double issue in one question: "Comment peut-[on] prétendre faire la preuve d'une absence d'archive?"[53] — how can we pretend to prove an absence of archive? The emphasis is on "prétendre" more than on the fact of giving proof of what is missing: the problem is not so much that we do not have this lost material, but that we might be tempted to pretend to give an exact

---

53   Derrida, *Mal d'archive* [42], p. 103.

account, or at least assessment, of what we do not in fact have. Derrida's point concerns primarily the epistemological perspective: the best we can do would then be to admit the coexistence, next to the traces we perceive, receive, and know of, of an abyss of absence. Of this submerged part of the iceberg (or is it the ocean itself?), we can only ever know imprecise things. Its silhouette forever remains impossible to draw, and with it, the draft of a reality past that we can never get close to, be it materially or speculatively. From this perspective, the more one tries to get close to a documental trace, to embrace the whole of an archive, the more one surrounds oneself with chimeras, and the less one is able to know something about it.

This does not only concern the individual position of a person wanting to consult, work on, or work with traces of the past, and who will necessarily be confronted with what one could call a dialectic of preservation and destruction that is inherent to archival processes. As institutions, archives are equally confronted by this challenge, which, beyond the fate of one singular document, is also reflected in the archiving process in general. All three instances — archives as an institution, archiving as a process, and the single piece of archive — face the dilemma of survival: what remains is incomplete, imperfect, something whose essence is not guaranteed to survive and that can only exist through an act of destruction. Derrida insists on the notion of "survival", emphasising the somehow organic life of what could otherwise be considered as an old piece of paper. Contradicting archivist Terry Cook who argues that the "twin pillars of the archival profession, appraisal and arrangement/discussion"[54] should be front and centre in this debate, other philosophers desert all form of archival agency to embrace this dialectic by shifting agency to the social body.

Foucault situates it in discursivity, engaging culture at large.[55] What interests Foucault in one relevant passage of *Archeology of Knowledge*

---

54   See Cook, *What is Past is Prologue* [38], p. 20.
55   Foucault states the following in *Archéologie du savoir* [53], p. 177: "Par ce terme [= archive; A.B.], je n'entends pas la somme de tous les textes qu'une culture a gardés par-devers elle comme documents de son propre passé, ou comme témoignage de son identité maintenue; je n'entends pas non plus les institutions qui, dans une société donnée, permettent d'enregistrer et de conserver les discours dont on veut garder la mémoire et maintenir la libre disposition. C'est plutôt, c'est au contraire ce qui fait que tant de choses dites, par tant d'hommes depuis tant de millénaires, n'ont pas surgi selon les seules lois de la pensée, ou d'après le seul jeu des circonstances, qu'elles ne sont pas simplement la signalisation, au niveau des performances verbales, de ce

is neither the archives kept by a culture to document its identity, nor
the institutions that record and preserve texts that are to be kept and
made available. His focus is on how the discourses deployed in these
contexts have emerged as verbal performances arising from an order of
the mind and an order of things, as the result of relationships at play.
In a way, Foucault and Derrida present similar views of archiving as
an informing order: it is a selection process in which a material trace is
not only preserved, and therefore destroyed in some of its dimensions,
but also gains a place in the construction of a collective memory as a
discourse, and an awareness of human limitation. This order of things is
not established once and for all in the archive: archiving remains a process,
and an archival piece keeps changing shape, shifting and evolving as
time goes by. The complexity of the object and of the processes is in many
ways an incitement to open up metaphorical spaces.

Barthes uses a positive image in *Writing Degree Zero* while delving
into the processes of the making of text, namely that of the magic ink.[56]
Any written trace reveals itself over time, with new layers appearing that
display new and always denser connections to the past. As opposed to
the notion of a permanent self-destruction, the magic ink suggests that
the ongoing process makes way for new layers of meaning to appear over
time. Rather than an impoverishment, the work of time is an enrichment:
history as an accumulation of temporalities is inscribed in the new layers
of meaning that appear, and not separated from them.

This idea of a fusion, or a coexistence of various, contradictory drives
that all linger in the archival document — the trace — is famously illustrated by Freud in a passage of *Civilisation and its Discontents*, where he
describes the impressions made by a walk through Roman ruins. The
same metaphor used by Freud to describe the structure of human soul can
serve just as well to describe the complexity of the archiving process.[57] At

---

qui a pu se dérouler dans l'ordre de l'esprit ou dans l'ordre des choses; mais qu'elles
sont apparues grâce à tout un jeu de relations qui caractérisent en propre le niveau
discursif."

56   See Barthes, *Le Degré Zéro de l'écriture* [22], p. 20: "Toute trace écrite se précipite comme
un élément chimique d'abord transparent, innocent et neutre, dans lequel la simple
durée fait peu à peu apparaître tout un passé en suspension, toute une cryptographie
de plus en plus dense."

57   See Freud, *Das Unbehagen in der Kultur* [55], pp. 35-37: "Wir greifen etwa die Entwicklung der ewigen Stadt als Beispiel auf. Historiker belehren uns, das älteste Rom
war die Roma quadrata, eine umzäunte Ansiedlung auf dem Palatin. Dann folgte die

first, Freud recapitulates the historical evolution of Rome following traditional historiography: Roma quadrata, Septimontium, Republic, Early Empire. He then notes that what he sees while wandering through the city are ruins, but of what has been rebuilt over the original buildings after fire and destruction. All remnants of Ancient Rome appear entangled in present, Renaissance and modern Rome, while many ancient remnants remain buried below the ground.

"Ruinen, aber nicht ihrer selbst" — ruins, but not of themselves, writes Freud: what is decaying here is not the original, but what modernity has made of old times ("nicht ihrer selbst, sondern ihrer Erneuerungen aus späteren Zeiten"). What was once buried is re-emerging, amidst the hustle and bustle of the present city ("in das Gewirre der Großstadt"). All the different layers of the city's history are intertwined, not to be dissociated from each other. The vocabulary used by Freud is not one that suggests an entirely organic growth of old and new together. It evokes moments of violence in both directions: violence imposed by the modern on the old ("Brände und Zerstörungen") as well as an unwanted emergence of the old in the new ("Einsprengungen" as an unexpected disturbance). In the lines that follow, Freud expresses a form of distress: the real Rome he is visiting is nothing at all like what he had read about it beforehand. It consists of an accumulation of disorderly traces in which it is impossible to distinguish what is antique from what is fake, what is old from what is new. Rome and its thousand intertwined layers call for some kind of order if you want to gain orientation. But Freud knows that the order he would need is not necessarily the same as what someone else would wish for. Any definition of order is bound by the requirements of one's own temporality. And it is, in fact, tempting to let nature have its way and let plants grow on old stones as they see fit, to let the natural decaying process take place undisturbed by the wish to impose rational

---

Phase des Septimontium, eine Vereinigung der Niederlassungen auf den einzelnen Hügeln, darauf die Stadt, die durch die Servianische Mauer begrenzt wurde, und noch später, nach all den Umwandlungen der republikanischen und der frühen Kaiserzeit die Stadt, die Kaiser Aurelianus durch seine Mauern umschloß. [...] Was jetzt diese Stellen einnimmt, sind Ruinen, aber nicht ihrer selbst, sondern ihrer Erneuerungen aus späteren Zeiten nach Bränden und Zerstörungen. Es bedarf kaum noch einer besonderen Erwähnung, daß alle diese Überreste des alten Roms als Einsprengungen in das Gewirre einer Großstadt aus den letzten Jahrhunderten seit der Renaissance erscheinen. Manches Alte ist gewiß noch im Boden der Stadt oder unter ihren modernen Bauwerken begraben."

recording and preservation strategies that are anchored in a time and space, and which are all the harder to define as one stands amidst it.

Applying this perspective to text as a trace of times past can be enlightening in many ways, revealing the tension between cultural constructs and natural evolutions of materiality in general. Gaining clarity on origin, transmission and perspective plays as much of a role in the archiving process as ensuring the right humidity, light and temperature: one needs both the one and the other, as well as awareness of the fact that neither of them will provide conditions that will make it possible to keep any documentary trace unchanged forever, nor to keep all of them. Change it must, and change it will.

Archive is a moving order of traces, caught in an ongoing process of self-definition. Distinguishing what happens to the single documentary trace and to archives as an institution does not make much difference: what matters in both cases is the archiving process. This process is a dynamic dialectics of preservation and destruction, destruction and preservation, that affects any trace of the past we might be tempted to interpret, including texts. Freud also points to the difference between the transmission discourse and the image that the city of Rome gives itself. The question is not about assigning blame to either the architects who covered ruins in concrete or the historians who describe what might be an imaginary city of the past: it is about accepting that we will never know what Ancient Rome looked like, and still make sense of Rome and of what historians and novelists write about it. Rome will always be Rome, and always be another Rome.

In this chapter, I have considered access to text from the point of view of archives. Although they have been in use for a long time, archival processes still challenge our vision of the past. In the archival profession, the increase in document mass in the 20th century, then the digitisation of records and sources in the late 20th century, have led to a shift in practices and the development of theories that encompass a more complex media and take into account the wider socio-cultural relevance of archiving. Philosophers, on the other hand, have come to challenge the notion of origin at the core of archival evidence. But both perspectives recognise the key role of archives as documents, as collections and as institutions in constructing discursivities throughout our history.

Archives provide structured access to text, and they have been doing so for a long time, alongside political evolutions. Taking them for granted is a cultural privilege, but one that should not prevent us from trying to understand underlying techniques (increasingly complex in the digital context) and epistemological challenges (such as the status of traces from the past), while making use of this privilege.

2. Publishing, editing, and their digital transformation

Archiving is only one process through which text is made available for reading. It is comparatively old and relies on technologies developed several centuries ago and applied to any form of portable writing. Since Gutenberg, printing devices have made wider dissemination in a print form possible. Text has become easier to carry around, and its duplication has contributed to a major increase in text mass.[1]

While archiving is about preserving, recording, and making available unique documents in one place, publishing is about multiplying versions of the same textual content, which is then distributed in an array of places, to a wide variety of readers. The symbolic as well as economic value of a single printed specimen is different from that of the unique archive: it is but one exemplar among many that are all similar to one another. The economic trade-off underlying print and the dissemination of its byproducts is more complex than that of archives. One of the reasons for this complexity is that more actors are involved in the dissemination process: a writer, of course, but also, at the very least, a printer, most likely also a publisher, a typesetter, copyeditors and critics. Publishing print products has been from the outset a capitalistic activity. Books, considered here as the epitome of circulating print,[2] have an economic value as objects of trade, and they have a cultural value. The economic dimension and the cultural one are closely intertwined.

Compared to the confines of the archive, the transformative value of publishing lies in its distributing capacity. Historically, print made it possible to duplicate and disseminate a text widely, which archiving did not. Printing, and even more so publishing, involves a range of actors and techniques and is more organically integrated in cultural practices. In

---

1   See Mac Luhan, *the gutenberg galaxy* [78].
2   Book historians (see for instance Barbier, *Trois cents ans de librairie* [21] and Wittmann, *Geschichte des Deutschen Buchhandels* [104]) use designations that embrace this dimension ("librairie", "Buchhandel").

      https://doi.org/10.11647/OBP.0355.02

the context of western cultures, at least, one generally reads more books than one comes in contact with manuscripts.

In the following chapter, I explore how publication as a practice of text dissemination adds to the use of archiving to make text accessible in a modern socio-cultural context. While the previous chapter considered text in general, in this chapter I focus more on literary texts — not only for aesthetic reasons, but mostly because they concentrate questions of authorship and make it possible to illustrate a wide array of questions pertaining to access to text.

The first section is focused on the shifts brought about by the function and figure of the publisher in the European context. I will show how publishers became key actors during the 18th and 19th centuries, essential to the development of literature as we know it today. I delve into some detailed analyses, highlighting the relevance of the relationships between writers and publishers, and shedding light on the mechanisms at the core of modern literature as a process of transforming a text into a work of art or even an œuvre. In the second section, I will bridge the gap to digitisation processes and explore what digitisation means for the type of text representations involved in publishing and archiving as two cultural practices of text dissemination and preservation.[3] Looking into the digitisation of modern textual heritage, I argue that the changes created by the digital medium do not concern the quality, especially the media quality, of our approach to text, as much as they concern quantity.

2.1 From text to book to œuvre

Traditional archiving as I have presented it is set to preserve the primary materiality or media quality of the text in question. In the case of the old family papers found in a drawer or of the postcard collection that can be purchased at a flea market, it is the manuscript that is preserved in an archive — the postcard itself, or the correspondence where the pages are in a specific order — and not simply a transcription of their textual content on another device. As already discussed, this approach has the disadvantage of exposing the unique copy to decay, and with it the disappearance of material and textual content. Another way of preserving

---

3   From a Human-Computer Interface perspective, see also Feinberg, *Beyond Digital and Physical Objects* [49].

text content is to duplicate it, which is usually done with the purpose of augmenting its impact, especially by reaching a wider audience. In this context, preservation is paired with dissemination. The publication process (making text available to an audience, making it public) involves additional layers of editing to the archiving process. The preparation of a text for publication follows different rules, sometimes complementary to those of archiving, and involves other actors, that I now present in some more detail.[4] These observations emanate from analyses rooted in a specific context: that of modern textuality. I will set aside all digital notions in this first historical section, turning to digitisation processes only in a second step.

I will begin with turning back again to the old papers found in your family home. I will assume the sheet bundle looks like the manuscript of a novel, and consider that the great-great-great-grand parent who wrote it not only preserved the manuscript for you to eventually find it at the bottom of the drawer, but for a wider audience to read it. Let us assume that your hypothetical ancestor would have wanted to publish this novel.

Following the description I gave in section 1.2.1, I would say that here I am considering texts originally produced by a particular hand (maybe helped by other hands, such as a copyist's), and intended for a wider audience. Reaching this goal is facilitated by printing techniques that transform the medium of the text in such a way that it becomes easier to access for a larger readership. The transformative value thus achieved is reflected in the marketplace at large (what is the price of a book? what is its relevance in the concerned economic structure? what is the number of books necessary to achieve such a relevance?): a book has a place in a state's economy or even in a global one. On a cultural level, it affects reputation mechanisms that contribute to establishing a hierarchy in the types of texts that are circulating at a given period in a given area.

Such a hierarchical approach prevails in literary studies. A random text has no specific value on this scale until it has established its literary market value. A work (of art) in the form of a printed and distributed book deserves more reverence than a simple text, as it marks a greater

---

4   What I am basing the following argumentation on is a very general presentation. While bringing complex mechanisms down to a generic description bears the risk of over-simplification, it facilitates an interdisciplinary perspective that encompasses a variety of aspects, as Darnton himself argues in his seminal *What is the History of Books?* [39].

degree of achievement and materialises a well-identified economic value and cultural capital. An œuvre consists of several works by an author that, considered together, have achieved a more superior status. In the context of an œuvre, even lesser texts (drafts manuscripts, correspondence — textual material with no primary market value) gain new significance.

Before the golden era of digital self-publishing, writers had seldom been in a socio-economic position from which they could themselves supply publications of their texts, by presenting them in such a way that a wider audience would have had access to them in exchange for payment. Writers in search of fame and money could hardly rely on a business model of that type. Publishing requires funds and skills in order to provide a context favourable to dissemination and to selling a literary work on a larger scale. This is where the publisher comes into play. Considering the complex relationships between writers and publishers also provides a shift of perspective in the history of literary text production.[5]

I will first analyse the mechanisms at play in the relationships between a writer and the actors that actively intervene in shaping a book for publication, focusing mainly on the publisher. Then I will then present two case studies based on early 19th-century German literature before I turn to questions concerning the digitisation of published textual heritage.

2.1.1   The deal between writer and publisher

Published texts whose superior literary quality is recognised are considered as works of art, and, taken as a whole, as lifetime œuvres. Not every text is part of an œuvre, but every literary œuvre is at its core composed of texts. The mechanisms underlying the transformation from text to book to œuvre, and the role of media transformation in book form, are not trivial. In the same manner that I worked out different stages of text constitution in section 1.2.1, l will now delineate the different stages in the evolution of a text transforming into an œuvre in an early modern context. The two temporalities of the evolving manuscript and the nascent book and their different stages, as we shall see, are partly intertwined.

Print applies to a range of formats: posters, journals, pamphlets, chapbooks, books, and more. Although books have not been the primary

---

5   See also, in the perspective of social history, Schmidt, *Die Selbstorganisation des Sozial-systems Literatur im 18. Jahrhundert* [95].

print format to disseminate even literary content for a particularly long time,[6] I will talk of "books" here, not because I care only for books in the strict sense, but because the notion of a book encapsulates the cultural capital that is at stake in the media transformation from manuscript to disseminated print, especially in the case of literature.[7]

The process of publication can be schematically described as follows: a writer conceives and writes a text, then entrusts it to the publisher who brings this textual content to book form, duplicates it and hands it over to readers in exchange for money (via distributors such as booksellers), a part of which is then paid to the writer.[8]

In the first step of this process then, the writer conceives and writes down the text. As I have shown in section 1.2.1, realising this first step can take some time and effort. It is not always achieved at once, but more often than not requires drafting, re-writing, and a first copying phase that applies solely to the manuscript, not to the print version ("copy" has different meaning in the publication process, depending on whether it is applied to a manuscript or a print)[9]. You can still consider it as one step, one creative phase that includes all events happening between the moment when the writer has their idea for the text they want to write and them sending a text out to a publisher: the draft, the first clean version, copies, and edited copies are part of this process, which has a high potential for alterations of all sorts between its beginning (the idea) and its end (sending out a copy to the publisher). Some authors prefer to dispose of the drafts altogether, others write clean drafts, others dictate them. There is not one unique way to write; there are rather almost as many ways as there are writers.

---

6   The printed press, with its feuilleton format for novels (in newspapers or magazines), was much more affordable than books and hence more popular. Only when the price of books could be significantly lowered, and literary publishers strove to reach a wider audience with high literature, did books become more popular. In Germany, the role of early 20th century *Kulturverleger* such as Samuel Fischer was key in that process. See Kuhbandner, *Unternehmer* [73].

7   I am thus following the discipline concerned with the dissemination of print culture, which is called book history.

8   The schema proposed by Darnton includes more actors [39]. I have chosen to simplify a fundamentally complex and changing set of relations, in order to provide general orientation. It does not account for historical and cultural variations and is strongly rooted in late 18th-century mechanisms when the publishing profession established itself.

9   A copy of a manuscript is usually supplied by hand; a copy of a print is another print.

Whatever the preferred procedure, providing text is seldom a solitary activity, even before it is sent out to the publisher. It can happen that the writer does not disclose their writing activity to anyone else.[10] But I will assume that other people are involved (family or friends as test readers, for instance), as is usually the case. It can even happen that the publisher is already part of this process already. In the case of book projects by authors who are already in contact with the publisher with whom they wish to work on a specific book project, authors and publishers can discuss structure or content — even if it is only the number of pages envisioned — which is itself bound to have an impact on the final text. There might even be yet another person involved. Texts the publisher receives may have been copied by the author personally, or by a copyist. In the ongoing negotiation between writer and publisher regarding content, the intervention of the copyist provides room for assigning this third party responsibility in delicate cases.

A second step in the realisation of the print version of the text follows this first one of conception and redaction. On this second step, the writer has conceived and written the text, and now entrusts it to the publisher in order for them to bring this content into book form. This step too involves not just one, but a series of actions. The text received by the publisher is read and edited, at the very least in terms of typographical errors and obvious mistakes, possibly even with more substantial interventions. Additionally, the transformation process from a writer's copy into a typeset print template involves decisions regarding the font that will be chosen and its size, type of paper, and book format: all things necessary to generate a first printed version of the text. The publisher can be supported in this decision-making process by various professionals: printers, copyeditors, typesetters. Depending on the requirements in each case, copyeditors may check mainly for typos, or their intervention may extend to much more consequential parts of the text. Moral and/or political censorship remained the rule for a long period of time in the modern

---

10   The weight of social conventions is not to be underestimated there. Depending on period and social context, it could (and still can) be dangerous to be outed as a writer. Many women who published anonymously in contexts where it was frowned upon for people of their sex were likely to send their manuscripts directly to publishers, without showing them to anyone beforehand, even more so than men who were struggling to make a living as writers and had to produce texts under so much pressure that there was no time left for other input.

era and framed the copyediting process; I will come back to the question of censorship in some more detail below. Here, let me simply say that there is obviously already a likelihood that the publisher will need to get back to the writer once these interventions are realised. The typesetter's intervention is definitely one that leads to discussions with the writer too, if the writer is in a position to intervene at this point. Writers usually have precise expectations when it comes to the form of their work: is the font elegant enough, the paper white enough, the binding (where there is one) soft enough? Many a material question is to be discussed in the process of shaping a book.

It is technically possible for the publisher to not consult at all with the writer, but this is rarely the case in the next step of the publishing process, which delivers the first version of the book, the so-called proofs. This first printed version of the text is provided for the author to check and correct. Proofs are printed in only one copy, with the sole goal of checking both printing techniques (layout, typesetting, etc.) and content (correct reproduction of the writer's handwritten or typewritten copy from step one). In general, at least one round of proofreading is planned for each publication, sometimes more. Carefully checking proofs is a delicate process requiring authors to immerse again in a text they might have entrusted to the publisher months before. If the first proofreading raises major issues, such as the insertion of several pages in the wrong place in the book, or anything that would massively disrupt the structure of the printed book, then a second, improved iteration of proofs has to be proofread as well.

A publisher's greatest wish, when it comes to proofs, is for the author to make as few modifications as possible. Even a change in a syllable or a word can lead to a shift in the page that will disrupt the layout, the placement of chapter beginnings on uneven pages, for example. When an author re-writes a text completely, based on the proofs, the publisher may have to start the entire typesetting and printing process all over again.[11] This costs not only time, but also money: the publisher then needs to spend considerably more resources in the book-making process. Every modification is costly, and to go over several sets of proofs has a notable

---

11   Some famous authors are known for completely redrafting proofs, to their publisher's despair. See for instance this digital version of the proofs of Baudelaire's *Flowers of Evil*: `https://gallica.bnf.fr/ark:/12148/btv1b86108314/f23.item`.

financial impact. In other words, this third step also provides room for negotiation of a balance between text accuracy and financial cost.

Even in this simplified description of the publishing process, you can see that the publisher's role is not limited to spotting typos. The publisher has a range of opportunities to contribute in-depth to the form and content of what they publish — room for intervention that can be interpreted as a form of authority over the text. Authors from the 19th century, such as Heine, have claimed that publishers' interventions equate to censorship.[12] In this case, the context was rather tense both in terms of the political situation and of the critical content of the writings. But more generally, if a publisher prevents a writer from using a specific, controversial wording, it can pass for censorship, whether it is officially branded as such or not. However, if an author does not even use a specific wording in the first place because they know that it would be frowned upon, and instead submit a manuscript to the publisher that avoids controversial wording altogether, should that be called self-censorship? Where does adjustment stop and censoring begin? It is in fact extremely tempting to brand as censorship any external element that leads to a modification of the text under some kind of constraint — but then, from this perspective, a large part of the production of a text would be considered to be modelled by censorship.

A text is always created in a specific socio-political context, with its moral, and often religious constraints. I prefer to use the term "censorship" in cases where an external organ is active in controlling the content of what is published. In 18th- and 19th-century German literature, this concerns mainly political censorship, but also religious censorship. Both church and state had appointed personnel to monitor the enforcement of the rules they had imposed on behaviour but also publications. Actual censors were commissioned, and publishers had to submit to them everything they intended to publish.[13] This opened the door to a parallel, clandestine book market that would not be subject to official censorship rules. Such literature followed slightly different publications mechanisms.[14]

---

12   On the complex relationships between Heine and his publisher Campe, and the way Campe navigated between official censorship and Heine's sense of what he could print, see Ziegler, *Julius Campe* [105].

13   On censorships mechanisms, see Kiesel and Münch, *Gesellschaft und Literatur* [72].

14   For more information on this topic, see the numerous excellent publications by Robert Darnton, for instance his *Censors at Work* [40].

There is still one final step to mention in the process of negotiating text between writer and publisher, which relates to the final duplication of the text for readers in return for money, a proportion going to the author. While the actual quality of the book accounts generally for its success, it is also true that this success depends on the reputation of both publisher and author. If an author is already well-known, their production will attract buyers. Similarly, a famous publisher will have an established audience. An author's name or a publisher's name can serve as a selling point.

In the case of an established author, it is the publisher's mission to capitalise on their existing reputation. Print distribution will have to be well measured, contact with booksellers optimised, and advertisements targeted, in order to increase the benefit in relation to what the publisher invested into the production of the book.

If one looks at the way a book presents itself to its readers, author and publisher are traditionally both involved. When I open a book, what I see on the title page is the author's name, the book title, maybe a place and date, but also, even if it is in smaller print and somewhat lower on the page, the name of the publishing house. This is the face of the book, delivering its identity to its audience.

Now to the distribution process itself. Readers are essential to this final step in the book production. Readers (or rather, what author and publisher imagine them to be) are the horizon of the whole negotiation journey in each of the previous steps. The concept author and publisher have of the book's audience becomes vital in the distribution phase. It can well be that they have very different expectations, but in order for them to come to an agreement, a minimal overlap in their respective sense of what the reception of the book should look like will be essential.

In his *History of Reading*, Alberto Manguel writes:

> Almost everywhere, the community of readers has an ambiguous reputation that comes from its acquired authority and perceived power.[15]

The focus is on the respect due to readers, a reverence that can be explained by the fact that the dynamics of any audience is unpredictable.

---

15   Manguel, *A History of Reading* [80], p. 35.

The success of a book depends on audience reaction, and that reaction can only be channelled to a certain extent. Some authors consider that readers do not understand their work (mostly in cases where they have had limited success), and they insist on readers being wrong and the book being right. This measure of self-protection does not always make them want to publish less, however. But then again, the financial incentive of publishing books does not come from people reading books, but from people buying them.

This is, however, only the pecuniary point of view. In terms of reputation (and financial value feeds on reputation), the goal is not solely that readers buy a book, but that they encourage others to do so in order to multiply sales. A good reader in these terms is someone who makes others want to read too. Reviews, for instance in journals, can be a major asset in disseminating the reputation of a book and encouraging more people to buy it, but hearsay might be even more efficient. And literary critics oftentimes have a very different opinion from that of a wider audience.

Financial success and literary reputation do not always work in perfect combination. Commercial literature can attract a wider audience than high quality literature even if, in the end, it is the high literature that will pass into literary history and keep being read beyond its author's lifetime.

There is not one simple unequivocal way to define good literature. Is it that which receives good reviews, that which is awarded prizes, that which sells well and provides its author financial security — or that which will still sell 200 years later? These recognition processes are long and complex.[16] Trying to disentangle them sheds light on how arbitrary judgements can become authority arguments if they align with socio-cultural values of a specific time and place.

In all stages of the preparation of a text for publication, intellectual production and financial issues are intertwined to the point that they cannot be considered independently from one another. Author and publisher play interconnected roles from the creation to the dissemination of a book. Depending on the period and the situation of the literary market, their relationships operate differently in the process of transforming a

---

16   In Germany, the so-called canon debate is a topic of scholarly, didactic and even public interest, regularly discussed in the press. For a systematic approach, see for instance Freise, *Literaturwissenschaftliche Theorien und Modelle der Kanonbildung* [54].

text into a book, and ultimately an œuvre. To some extent, working out a literary history based on the relationships between writers and publishers would also mean writing a new history of the book market. And it would shift focuses in traditional literary history, as I would like to show to close this section.

This narrative has its starting point in the late 18th century. Before that period, the book market was based on a barter trade that involved, apart from writers, printers and booksellers. The profession of publisher — the one person who incarnates the interface between printer and bookseller — emerged in the late 18th century together with the professionalisation of the writer job. Professionals who acted as publishers, that is, those who started investing capital in publications, were still often called printers or booksellers in the late 18th century.[17] The emergence of the profession of publisher does not mean that all the other professions involved in the production of prints were at once strictly separated from each other. In fact, you could find publishers who were also writers, writers who had copyeditor jobs, copyeditors who translated and printed, publishers who were also booksellers, etc.

In the late 18th century, the expansion of the book market and its financial potential led to various abuses. Printing technologies had become cheaper, literacy was on the raise, and the book market promised to become an interesting source of income for professionals in the printing branch. All scholars were affected by this evolution. In his 1797 *Metaphysical Foundations*, Kant took a look at the book as a cultural artefact of primary relevance.[18] In this text, he focuses on the book as a medium of transmission of text and positions first the writer (or author), then the publisher as actors in its transmission to an audience. His argumentation revolves around the materiality of the book itself insofar as it embodies the text. Three key actors are involved, and are situated in relation to this

---

17   See Wittmann, *Geschichte des Buchhandels* for this and the following historical overview [104].

18   See Immanuel Kant, *Metaphysische Anfangsgründe*, 1797: "*Was ist ein Buch?* Ein Buch ist eine Schrift (ob mit der Feder oder durch Typen, auf wenig oder viel Blättern verzeichnet, ist hier gleichgültig), welche eine Rede vorstellt, die jemand durch sichtbare Sprachzeichen an das Publikum hält. – Der, welcher zu diesem in seinem eigenen Namen *spricht*, heißt der *Schriftsteller* (autor). Der, welcher durch eine Schrift im Namen eines anderen (des Autors) öffentlich redet, ist der Verleger. Dieser, wenn er es mit jenes seiner Erlaubnis tut, ist der rechtmäßige; tut er es aber ohne dieselbe, der unrechtmäßige Verleger, d.i. der *Nachdrucker*", quoted from *Gelehrsamkeit ein Handwerk?* [92], p. 230.

material artefact. The writer is the one-time producer of the language-based signs that make up the text; the one who presents this text to a wider audience in the name of the writer is the publisher in the case where the duplication is legal, the reproducer in the case where it is illegal. Kant points here to a major debate of the late 18th century, which eventually led to the first (albeit comparatively late) copyright rulings in the German-speaking area. In the absence of an international legislation on reproduction rights for intellectual works, several printers set themselves up to reproduce books without having any kind of agreement with their authors. In this case, the author did not receive any financial compensation for the printer to reproduce their work. This dysfunctional *modus operandi* was partially corrected in some German states and regulated in others at the beginning of the 19th century, which contributed to establishing the reputation of specific states and cities as publishing hubs based on their legislations and how respectful they were of writers' financial and creative aspirations.

Questions of reputation have a different impact for authors and for publishers, which also depends on the type of book that is being published. My focus here is on what is called high literature, although I know very well that it made up for only a small part of the book market in the German-speaking area of the late 18th to early 19th century. Literature, just like any other published work, is dependent on the conditions of the book market in general. But, maybe more than any other domain in the book market, high literature is one that involves strong identification mechanisms with the published work, at any rate more than might be the case with more technical productions, such as household-related guidebooks that were a much more widely disseminated genre than high literature at the time. Strong identification with the text also led to more emotional responses to the money issues at stake in the process of producing a book. This was true for both writers and publishers, and added to the complexity of their relationship.

How authors and publishers interact can vary. The agreement between the two parties can be more or less formalised: in late 18th-century German-speaking areas, most authors, and most publishers, operated without any form of written contract. This left the question of the shared responsibilities open to some extent, and could lead to an unresolved situation between author and publisher. Who is in charge of deciding on typesetting, on the number of copies, even on the wording, if this is not settled in a contract? The work, in the form of a book, is a common endeavour of both writer and publisher, requiring acknowledgement of what they share and how they share it.

Even in the cases where a formal contract exists, it does not necessarily remove all ambiguities, if only because things change with time. Contracts could very well deal with not simply the production of one book, but several or even all of them. If a publisher wanted to keep working with an author (and vice versa), specific agreements could bind them for more than one publication project. Hence the long-time associations of authors and publishers that become an intellectual, cultural, and economic tandem— and sometimes also the estrangement that becomes gossip for the cultural elite.

If the reputation or success of one of the agreement partners evolves differently from that of the other, an initially stable and clear relationship can be disrupted. But it can also be that the emotional investment of the partners varies over time. A particularly close relationship between a publisher and an author could be qualified as a "friendship" by either one or both of them at a certain point in their relationship. This choice of word to designate a work agreement involving money negotiation and transactions opens the door to a wide range of potential misunderstandings.[19]

Does that mean that the relationship between author and publisher consists of so much more than business that it cannot be embraced by a contract? Therein lies essentially the contradiction. Reciprocal trust is at the core of their relationship, and the business model itself requires both actors to identify themselves with each other: the author has to feel at home in the publishing house in order to be able to entrust their text to the publisher; the publisher has to see the text rooted in the publishing house's profile at the very least, if not in their own intellectual identity, for it to be published.[20] This leads to a sense of reciprocal identification effects, but also to possible conflicts of interest, disputes on competence areas, and more. In a way, authors have to separate from their text and its uniqueness in the moment when they agree to a contract, be it virtual or

---

19   See Fischer, *Merkwürdige Verbindung* [52].
20   In *Der deutsche Buchhandel* [67], Hiller says, in that sense, "Autor und Verleger leben in einer Art Symbiose" — author and publisher live in some sort of symbiosis (p. 77).

real, with a publisher. At this point, the publisher takes possession of the text to some extent. The artefact at the core of the transaction is charged with emotionality, not only symbolically, but almost essentially. As a result, the relationship between author and publisher is impregnated by authorship issues that no contract, however extensive it may be, can fully resolve.

Imagining a history of literature that focuses on the relationships between writers and publishers would make it possible to assess authorial processes in the production of text in the form of a book as a work of art, and in the process of creating an œuvre. It is also a way to consider the media transformation of the text as a process that fully determines how readers will access it. The publication process I have described above, although presented as a general framework, best applies to (early) modern literature. In order to make the transformative impact of the relationships between author and publisher at text level easier to grasp, I next elaborate on a few examples from the German context . From there, I will then move to the media transformation involved by digitisation processes and how they, too, generate shifts in access to text and its constitution.

2.1.2   Negotiating the œuvre

At first sight, relationships between authors and publishers appear to be a collaborative process that can be framed and regulated in a productive manner. But productivity comes at a price, and it is worth looking behind the curtain when it comes to transforming literary texts into valuable works and ultimately into a recognised œuvre. My two examples from the late 18th and early 19th century, Goethe and Tieck, shed light on different strategies.

Johann Wolfgang von Goethe and Ludwig Tieck followed different career paths, but they share a few biographical features. They are both considered major German writers, and they lived well into old age, Goethe died to the age of 83, Tieck 80, yet troubled for most of their life by a hypochondriac fear of dying. Their longevity did not bring reputation, money, and happiness alone: they both lost many dear friends in the course of their existence. In 1805, Goethe lost in Schiller his *alter ego*, and survived him for 28 years. In 1830, near the end of his life, the death of his son August deprived him of a strong asset in the management of his late publications, too. Tieck lost his friend Wackenroder in his youth, then his young-adulthood friend Novalis, and later his mature-adulthood friend Solger, who all died at a relatively young age. These losses affected their view on life, and their vision of their own work.

Their strategies with publishers were, however, quite different, largely because of their respective work ethics and ability to capitalise on past literary successes. Tieck played different publishers off against one another, while one of Goethe's strategies led him to focus ultimately on a single publisher — the most powerful one of his time, Johann Friedrich Cotta. Book historian Reinhart Wittmann considers Goethe as exemplary for the way in which relationships between authors and publishers evolved at the beginning of the 19th century, and I will draw on his argumentation in what follows.[21]

Goethe's literary career starts in 1765, when a bookseller from his hometown Frankfurt takes him to the Leipzig book fair (already then an institution, and still one today) and introduces him to publisher Philipp Erasmus Reich. A first volume of poems by the young artist was published anonymously in 1769, but Reich refused to print Goethe's next manuscript. Goethe then produced his next works by self-publishing them, including his epic drama *Götz von Berlichingen* in 1773, which turned out to be a success and was immediately reproduced in unauthorised copies by numerous publishers. These reproductions were sold without bringing the author any money. Nonetheless, he had succeeded in gaining traction within the literary arena. For his next opus, an epistolary novel entitled *The Sorrows of Young Werther*, he found a publisher and made enough profit to wipe out the debt left by his self-publishing phase.

From this moment on, he resolved to systematically request very high fees from his publisher for every new work he wanted to publish, without even showing them the manuscript. In addition, he negotiated with publishers via a third party: it was someone else (a person he trusted) who actually contacted publishers and negotiated the conditions dictated by Goethe. At this point in his life, he did not technically need the income from his publishing activity as he had been made a minister in Weimar. His hardcore negotiation tactics were less a struggle for survival than a

---

21   See Wittmann, *Geschichte des deutschen Buchhandels* [104], pp. 175-185.

symbolic gesture of revenge for the humiliation he had endured in his first years as a young writer.

Upon returning from a transforming trip to Italy in the late 1780s, Goethe tried to publish his first scientific work, the *Metamorphosis of Plants*, with publisher Georg Joachim Göschen, who had commissioned the first complete edition of his literary works. Göschen refused in somewhat crude terms.[22] It does not matter to me, Göschen said, whether Goethe wrote the book: a shopkeeper cannot be a philanthropist. Göschen's commercialism was hardly compatible with Goethe's intellectual aspirations, and even less with his self-esteem. A subsequent round of negotiations was undertaken with Johann Friedrich Unger, who had published a few of Goethe's books, but with so many printing mistakes that Goethe doubled his author fees. In order to compensate for his financial loss, Unger procured a parallel print (accordingly exempt of author fees), which he did not mention to Goethe. But Goethe noticed and refused to continue working with Unger. At the same time, Goethe had been discussing a much higher fee for his play *Hermann and Dorothea*, which was ultimately profitable to publisher Johann Friedrich Vieweg because he kept printing and selling it for years after the standard two-year contract had elapsed. This cat and mouse game is symptomatic of two circumstances: first, the lack of clear rules; second, the fact that both parties, author and publisher, were trying to negotiate financial advantage in relationships in which trust conflicted with the understanding each had of their own legitimate claims.

A glance at the discussions between Schiller and Goethe related to the publication of the journal *Die Horen* in the 1790s sheds light on cooperation with yet another publisher, Johann Friedrich Cotta.[23] The journal *Die Horen* was edited by Schiller and published by Cotta between 1794 and 1797, and became a major publication outlet of German Classicism. In a letter to Goethe from September 1794, Schiller underlines Cotta's "zeal and decisiveness", "tirelessly promoting the journal", and his "'punc-

---

22 "Ob ein Goethe das Buch geschrieben hat, ob es die höchste Geisteskraft erfordert hat, darauf kann ich als Kaufmann keine Rücksicht nehmen. Ein Krämer kann kein Mäcen sein", quoted by Wittmann, *Geschichte des deutschen Buchhandels* [104], p. 179.

23 Ironically, the correspondence between Schiller and Goethe, in which they extensively discussed negotiation tactics with Cotta, was published by Goethe with Cotta's publishing house a few decades later — magisterial proof of Cotta's ability to anchor his professional reputation on factors other than personal considerations.

tuality in delivering author and editor fees".[24] Schiller's praise of his publisher — aimed at convincing Goethe to contribute to the journal — was not only motivated by Cotta's qualities as a publisher, but also by the fact that Schiller needed the regular income provided by Cotta in exchange for the publication of the journal.

While it was Schiller's initial idea to publish contributions anonymously, Cotta insisted on having them signed by their authors, which would be a valuable selling point, considering these were Goethe, Schiller, Humboldt, Fichte — by then already known actors in the intellectual scene. Goethe refused categorically, pointing out that anonymity was the only way for him to remain free to write what he wanted.[25] Schiller and Cotta had to give up on their selling argument for the sake of Goethe's freedom.

Later discussions display a similar imbalance. Schiller writes to Goethe that Cotta wishes for more diverse, less abstract content.[26] Goethe's answer disregards the publisher's concerns and focuses solely on the relationship between author and audience. Let us proceed on our path — we know what we can deliver, writes Goethe, adding: I have known the farce of German authors for twenty years now, it just has to be played. The show must go on, so to speak: accommodating this perspective as well as Cotta's was an acrobatic manoeuvre that required all of Schiller's diplomatic and editorial skills.[27]

Around 1800, Goethe engaged in a long-lasting personal work relationship with Cotta.[28] It was Schiller who had motivated Cotta to try and publish Goethe's works. Schiller states that Goethe is too unique to have a standard market value: and since he is priceless, any price will

---

24   "Eifer und Entschlossenheit"; "unermüdete Tätigkeit in Verbreitung des Journals"; 'Pünktlichkeit im Bezahlen"; see *Schiller-Goethe-Briefwechsel* [96], p. 50.

25   Goethe writes: "Cotta mag recht haben, daß er *Namen* verlangt; er kennt das Publikum, das mehr auf den Stempel als den Gehalt sieht. Ich will daher den übrigen Mitarbeitern die Entscheidung wegen ihrer Beiträge völlig überlassen haben, nur was die meinigen betrifft, muß ich bitten, daß sie *sämtlich* anonym erscheinen; dadurch wird mir ganz allein möglich, mit Freiheit und Laune, bei meinen übrigen Verhältnissen, an Ihrem Journale teilnehmen zu können"; *Schiller-Goethe-Briefwechsel* [96], p. 73.

26   See *Schiller-Goethe-Briefwechsel* [96], pp. 103-104.

27   "Lassen Sie uns nur unsern Gang unverrückt fortgehen; wir wissen, was wir geben *können* und *wen* wir vor uns haben. Ich kenne das Possenspiel des deutschen Autorwesens schon zwanzig Jahre in- und auswendig; es muß nur *fortgespielt* werden, weiter ist dabei nichts zu sagen."; *Schiller-Goethe-Briefwechsel* [96], p. 105.

28   For more detail on the following analysis, see Fischer, *Cotta* [50], especially the chapter "Karlsbad und die Folgen".

be worth paying[29] — the argument could not be less oblique compared to Göschen's. Cotta's reaction shows a surprising display of restraint ("ich war zu schüchtern, in dieser Hinsicht etwas zu erwähnen" — I was too shy to suggest anything), explicitly asking Schiller to intervene in the negotiation as the middleman. In this first negotiation step, Cotta also points to the fact that this connection should not be a one-off.[30] The extent of Cotta's dedication to his star author was clear. Publishing Goethe's works turned out not to be profitable at all for Cotta. Goethe demanded extremely high fees and never really trusted Cotta. Evidently, Cotta deemed such a treatment tolerable in return for the prestige of calling himself Goethe's publisher.

After Schiller's death in 1805, negotiations between Cotta and Goethe became more complicated without Schiller's instrumental intervention, and ended up extremely troubled when it came to the production of the final edition of Goethe's complete works, the *Ausgabe letzter Hand*. The 1820s had seen the rise of new highly regarded publishers for works of literature, especially Georg Andreas Reimer and Friedrich Arnold Brockhaus, with whom Goethe had engaged in negotiations even though he was still under contract to Cotta. Goethe finally turned to Cotta in 1823 to discuss the ultimate edition of his œuvre, however. When Cotta answered with a request for additional information, this hurt Goethe's feelings.[31] What is more, the offer Cotta made for the 40-volume edition was lower than those the other publishers had already made. Goethe was irritated and did not react to his publisher's offer. In the absence of the now deceased Schiller as a middleman, Cotta commissioned art collector Sulpiz von Boisserée to enquire about Goethe's state of mind.

After Boisserée managed to sort things out and re-establish a constructive relationship between Goethe and Cotta, Goethe's son August (who was blinded by the — technically unrealistic — offers made by other

---

29   "Ein Mann wie Goethe, der in Jahrhunderten kaum einmal lebt, ist eine zu kostbare Akquisition, als daß man ihn nicht, um welchen Preis es auch sey, erkaufen sollte", quoted from Wittmann, *Geschichte des deutschen Buchhandels* [104], p. 181.

30   "Ich hege freilich immer den stolzen Wunsch, daß ein angefangenes Verhältnis der Art nie getrennt werden möchte, und ich werde daher immerhin das möglichste tun, es zu erhalten und diejenigen, die sich mit mir in solche Verbindung einlassen, es nie bereuen zu machen." — I have the proud wish that this bond shall never be broken, and I will do all that is in my power to maintain it: Goethe quoted from Fischer, *Cotta* [50], p. 113.

31   Fischer, *Cotta* [50], p. 668.

publishers) came across an edition of his father's *Faust* by Cotta, which the publisher had not mentioned to the author and for which Goethe had accordingly received no fees.[32] Cotta's argument that the rights he had acquired for the complete works included the right to reproduce single works that were part of them further worsened Goethe's mood. Taking advantage of the weaker position into which his son had manoeuvred Cotta, Goethe managed to negotiate even more profitable conditions, and the contract for the *Ausgabe letzter Hand* between Goethe and Cotta was signed in 1826.

This brief glimpse into a rich and complex context demonstrates that even in the case of major actors in the literary field, whose survival, life or career did not depend on the outcome of the negotiation, work relationships hang by a thread, and trust remains the answer.

The case of Tieck presents a somewhat different situation, but leads to a similar conclusion.[33] Tieck started his literary career by providing contributions for Enlightenment publisher Friedrich Nicolai's satirical collection, *Die Straußfedern* in 1794. Tieck was 21, and he continued working for Nicolai until 1798. The texts were published anonymously, without any title, as a series of numbered contributions. While the publisher was by then already famous, the writer was not only unknown to the public, but writing mostly on commission. He also translated French and English prose, which he did with the help of his sister Sophie and her husband, August Ferdinand Bernhardi. While Nicolai was well aware of young Tieck's literary talent, Tieck himself did not adhere to the satirical style of late Enlightenment *à la* Nicolai and this led to tensions. This period was one of the most productive in Tieck's life, however. He integrated these early texts into later editions of his works — subjecting them to a renewed editing each time, but keeping the basic stock.

Turning his back on Nicolai proved profitable for Tieck's literary ambitions. They ceased to work together after a quarrel regarding the publication rights of one of Tieck's works — a quarrel that became public, ensuring that the young author made a name for himself on the literary scene. Now he was young, famous, and a champion of true poetic inspiration. This prestige called for much higher fees than those he had

---

32   Fischer, *Cotta* [50], p. 671.
33   For the biographical part of this analysis, I draw on Roger Paulin's standard work *Ludwig Tieck. Eine literarische Biographie* [86] as well as on the chapter on "Tieck und seine Verleger" by Philipp Böttcher in the *Tieck-Handbuch* [27], pp. 148-164.

received from Nicolai. Tieck moved to Jena where he became part of the famous Romantic circle.

It was Friedrich Schlegel, his Jena roommate, who introduced him to publisher Georg Andreas Reimer in 1802 in the context of an edition of Novalis' works. Reimer had been active in Berlin as a bookseller since 1800, and selected his publications in relation to his political orientation, which, in the context of the Napoleonic wars, was decidedly German and patriotic. Reimer published both Novalis' works (which sold very well) and Tieck's (which sold modestly). At this point, Tieck had started repeating the same procedure he had experimented with when working with Nicolai: he asked for credit, again and again, for books that either did not sell well or which he never even delivered.

In the 1810s,[34] Tieck published his famous *Phantasus*, a collection of youth writings, comprehensively edited and framed with a narrative in the style of Boccaccio's *Decamerone*. He was financially supported by a patron — a former schoolmate who ensured his and his family's subsistence. Tieck kept asking for high fees from Reimer, but Reimer had seen through him and found a strategy by which he would not suffer excessive losses. Reimer always paid Tieck a little bit more of the advance fee he had asked for, in order to trick him into feeling guilty and actually writing and sending the promised texts. But on the other hand, he only paid him the fees due on sold volumes with lengthy delay, even years of deferral. All in all, Reimer did not engage in a highly risky business with Tieck.

Tieck started to notice the late fee delivery, and for several years, their correspondence stopped. Reimer kept sending money, but the reciprocal trust was damaged and, in the 1820s, two other publishers set out to win over Tieck and make him their house author: Josef Max and Friedrich Arnold Brockhaus. Tieck had by then settled in Dresden with his enlarged family, where he enjoyed a new creative phase. He tried to take advantage of the concurrence situation among the three publishers. In a first step, Reimer remained in charge: he took over the contract for Tieck's complete works (signed in 1827) but, in order to prevent further losses if Tieck did not deliver, he included the right to publish, five years after their first

---

34　On this period in Tieck's production, see Baillot, "Das Bild Shakespeares" [5] as well as Jochen Strobel in the chapter on "Dresden, Berlin und Potsdam" in the *Tieck-Handbuch* [97], pp. 108-111.

publication, all works by Tieck that had appeared in other publishing houses. Further, Tieck was to pay 100 thaler per month per manuscript for each one he failed to deliver on time. For the Shakespeare translation, Reimer opted for the opposite solution, and offered an additional 1500 thaler per volume delivered on time.

In 1838, Tieck wanted to break away from Reimer. This was extremely complicated as the contract was binding, tying him firmly to the publisher. He turned to the Brockhaus publishing house: from 1826, Tieck published novellas in their pocket book series, *Urania*. He received rather high fees for this work, but Brockhaus did not complain about the cost. He printed in small font, forcing Tieck to deliver more text content per page than was usual and in addition, the format was pre-defined and the author was compelled to stick to it. Tieck complained bitterly, saying it was "barbaric" not to be able to decide for himself where textual content would be positioned on line and page.[35]

The real crux in the relationship between Tieck and Brockhaus turned out to be Tieck's book collection.[36] Although he constantly lacked money, Tieck had managed to gather an impressive library of books over the years. But in 1839, he needed money so badly that he sold it to Brockhaus. It was really more a lease than a sale, as Brockhaus had agreed that Tieck could keep his books and receive a yearly amount of money in exchange for Brockhaus' right to dispose of the collection after Tieck's death.

As advantageous as the agreement was — especially as Brockhaus did not have any major financial plans for the book collection, which he only intended to eventually make available for research and teaching — Tieck still sold his book collection a second time, in 1849, to auctioneer Adolf Asher. Scandal could only be diverted with the help of a friend who bought Tieck's books back from Brockhaus in order for Asher to be able to proceed with the auction. Brockhaus' gesture of offering such a generous lease, an exceptionally friendly offer, was not honoured by Tieck. Here, again, trust appears to be a most fragile device.

Goethe and Tieck are only two examples, and from a time when there was little legal regulation available to frame the business relationship

---

35 See Böttcher, *Tieck und seine Verleger* [27], p. 160.
36 This discussion relies strongly on the chapter "Tiecks Bibliothek" by Achim Hölter in the *Tieck-Handbuch* [68]. See also the digital edition of the catalogue of Tieck's book collection at https://tieck-bibliothek.univie.ac.at/.

between authors and publishers. The loose legal framework of the early 19th century sharpened identification with the published work on the author's as well was on the publisher's side. These examples also show how unproductive it is to assume that there is only one authoritative version of any one work. Both Goethe and Tieck produced several editions of their works. The publishers influenced the output, but it was also authors themselves who kept re-drafting their works — based, inevitably, on the critical assessment of previously printed versions.

The relationships between writers and publishers shed a singular light on the history of literature. They do not tell a story of definitive versions and of authors' final words, but one of common writing processes. Manuscripts of literary works can be better understood when considered in the context of the correspondence between writers and their first reader circle, but also between writers and their publishers, as well as such commissioners as Boisserée in the case of Goethe's *Ausgabe letzter Hand*. All these contributing hands shaped the text into the work, then into the œuvre that we know today — because at the present point in the history of western literature, it is mostly famous works and famous authors that we read about.

But one could also look at this from the point of view of the history of print and book. What novel insights could be gained there, drawn from what material? One would not only need the correspondence between writers and their friends and colleagues to understand the shape in which prints have reached us. One would also need to systematically study book fair catalogues, censored drafts, account books of publishing houses, printers' correspondences, and accounts of overarching institutions such as the *Börsenverein des Deutschen Buchhandels* in the German context.[37] One would need publishers' correspondence to be fully integrated into the analysis of the creative process, maybe even that of their spouses, who often contributed a great deal to the family business. This is studied as a branch of history, but not in an encompassing manner as a branch of literary history.

---

37 The *Börsenverein* is an association that was founded in 1825 to secure the interests of publishers in the book market, and remains an active actor to this day. A wealth of scholarly works have been dedicated to its history; a short overview is provided on their website at `https://www.boersenverein.de/boersenverein/ueber-uns/geschichte/`.

The history of literature does not provide the tools to address this material, partly for pragmatical reasons. Archives of publishing houses are prone to destruction. On the one hand, it is a business in which the destruction of material is a question of survival: given a finite room to stock material, there is only so much that can be kept and, in general, it makes more sense to stock material that can be sold (such as recently produced journals and books) than what cannot (as is the case with archival material). On the other hand, especially for larger, established publishing houses willing to keep traces of their history, this destruction process is not always encouraged or facilitated. But publishing houses have been exposed to the most precarious conditions during the 20th century: they, and with them their archives, were specially targeted by shelling during both world wars.[38] Their strategic value was recognised, which endangered them in critical political contexts.

Literary critics at the time of publication too could be integrated much more explicitly into the transmission of literary texts than it is the case today. Considering that writers often redraft previous versions over time, the chances are that vocal critics, or particularly convincing ones, may have an impact on the editorial evolution of a text. Admittedly, such interventions are not easy to grasp. They require an excellent knowledge of the period, of its means of communication, and of influences: this is what frames literary quarrels and is where sensitive spots can be found. Critics often carry some responsibility in the posterity of a literary work, and hence contribute to the construction of literary history at large. The relationships between publishers and critics are another field in which one would be likely to find traces of the evolution of texts on their way to becoming œuvres, such as letters they exchange, and mandates they undertake. As I mentioned in section 1.2.2, the reception of a text is also part of a text genesis, just as text genesis is a reaction to the pre-existing reception of other texts or works. This intertextuality is at play in a variety of textual traces, and inherently questions the notion of a unique version of a text, as well as that of the unique authority over it. Placing the relationships between writers and publishers under closer inspection is one way to shed light on the many hands at work in a text, even when the writer is a famous, established author, and even without taking into

---

38   This is the case for the Insel Verlag; see Kuhbandner, *Unternehmer* [73] and Ziegler, *Buchfrauen* [106].

account all the ancillary work done at home by those who surround him (more seldom: her), and whose traces are even more volatile and difficult to grasp.

This information, gathered by scholars from various disciplines and with various interests in textuality, is brought to the attention of general readers through two major outlets. The first one is school material that provides an initial approach to literary works and to literary history, with varying levels of precision depending on the teaching level. Goethe's poems can be studied early in a school curriculum. In fact, both Goethe and Tieck wrote short poems that provide a simple access to poetry and that are used in classrooms, making them basics in German literature. Another essential way of channelling the reception of literary works are the editions of their works. Once authors are dead, these editions can be undertaken by family members, initiated by publishers, and/or be commissioned as scholarly endeavours. Editorial tradition is as old as print techniques, even older than them. Preserving and transmitting texts from Antiquity has been ongoing across centuries. It took on a new dimension in the early 19th century, when choices of modern authors to edit (who? how?) were not to be reduced to philosophical issues (should we follow Aristotle or Plato?), but steered by political orientations. Editorial traditions of the 19th century evolved differently even within Europe. Some emphasise text genesis, others context, yet others engage with the definition of a unique version of reference. Our knowledge of literature and literary history owes its key orientations to European editorial practices of the 19th and 20th centuries. This is where digitisation found us: struggling to cope with editorial traditions and to make sense of endeavours that had become scholarly or economic behemoths — 20 years or more for a team to edit the œuvre of one author in books so expensive that almost nobody can afford them.

In the next section, I will argue that the shifts brought about by digital media make it possible to reassess these orientations, and leads to a new convergence of archiving and publication practices.

2.2   Text quality from scan to digital edition

Much has happened between 1830 and the early 21st century. Political and social evolutions, shifts in economic leverage instruments and actors and, with the development of new forms of media, also novel ways to engage with readership.[39] I will not undertake a history of editorial practices since 1830, but rather take a big leap to examine what happens with the relationships between authors and publishers, and the hierarchy between text, work and œuvre, once text is made available online. In terms of textual materiality, I will remain focused on those literary works that are considered relevant heritage, as they tend to be included in digitisation programs and illustrate my argument incisively.

In the following section, I consider the way in which the shift to a digital environment modifies representations of text and discuss to what extent this leads to renewed forms of its reception. Does the fact that anyone can post anything online lead to a complete disintegration of the literary canon? Does the multiplicity of the forms of presentation available mean that we have stopped reading linearly and do not rely on the page structure anymore? I argue that the media change to digital formats has not modified our reading habits as drastically as we may think, and certainly not as much as might have been possible considering the extent of the structural change that has taken place. In the final section of this chapter, I discuss models of online access to text and explain why Open Access is susceptible to major leverage in the renewal of access to text, and more specifically to cultural heritage material. I begin by unrolling once again the process that has been presented so far: that of the constitution, editing, and publishing of text, but this time, in a digital setting.

2.2.1 Emancipation from the page?

For someone coming from traditional text or book studies, or even from the heritage domain, assessing online textuality is nothing short of intuitive. I first consider the different types of text that can be found online and how they can be transformed into genuinely textual formats. My goal is to delineate as clearly as possible why formats matter and what impact they have on access to text in general. Understanding the different representations of text available online is the initial step to acknowledging (and eventually applying) quality criteria to digital text.

---

39   See Mac Luhan's seminal *Understanding Media* [79], 1964.

While archives, libraries, academies, and publishers, to name only some of the key actors involved in text dissemination processes, were attributed roles with respective boundaries that crystallised over the past decades and centuries, the borders between these players and their activities are somewhat blurred in the digital context. For a layperson browsing the web, it is not easy to distinguish the difference between a webpage that is set up by a random individual, a site that contains valuable archived material, and a literary work of art. In order to understand the mechanisms that regroup such varied textual material under the overarching roof of digitality, I will consider it not only from the point of view of the user (how do I, as a reader, assess the type of material I am consulting?), but also from the point of view of a producer (how do I, as someone who wants to make textual material available online, transform text into a digital archive, or into a digital edition?).[40]

In the physical world — or rather in the analog world, since the digital is itself material — text presents itself on a page. It might be a manuscript page, a postcard, a typewritten document: as long as the page contains letters or even just signs, it can be considered a text. This is a piece of paper, a page on which legible characters are drawn or written, and form to some extent a coherent unit of sense.

In a digital setting, the same textual document can be accessed either in an image format (the scan of the manuscript), as a transcription of its content (the raw text), or in an annotated format with underlying information that is readable by the machine without being necessarily being visible to the human eye. Annotations can be added to images as well as raw text. An example of such an annotation would be, for instance, if abbreviations are expanded and additional information is supplied. As a reader, you would then know who the person mentioned in the manuscript is when the manuscript simply says "Mr R.", or in what century a text was written when the manuscript date reads "7 June 36". The additional information, whether names, dates or something else, is part of the metadata of the document concerned.

An image, a raw text, and a somewhat enriched (annotated) image or text are the most basic representations of text in a digital context. But

---

40   On the functional overlaps between reader, user and producer in the context of the web, see Rob Wilkie, *The Digital Condition* [102] and Elizabeth Bird, *Are We All Produsers Now?* [24].

one can also engage in the production of other, more complex representations of text.[41] In the case of literary works, statistical analysis of word occurrences can be calculated based on raw or annotated text and a visual representation of the result can be generated based on this calculation, for instance, in the form of graphs. The output, graph or otherwise, is then a new way of representing the text, one that includes the result of statistical exploration or analysis. Word clouds, displaying the words that are the most important statistically in a text in a font size relative to how often they occur in that text, are such a (rather simple) representation.

Following this method of calculating connections between elements of a text, you can represent networks of figures in a theatre play so that the relative importance of the relationships between them is represented visually. Such graphs are made available online, sometimes very prominently on a webpage, as a way for the reader to gain an overall visual impression, leading them, in a way, on a pre-determined path.[42] These representations of text require an additional level of abstraction as they are the result of visual choices, which are themselves derived from mathematical calculations based on the source text. But they can be considered as representations of text as well.

Yet, the difficulty lies not in considering that specific forms of visualisation are representations of text, but in knowing how to deal with them in terms of interpretation. How do you read a graph? And what does it tell you about the plot, the style, and the characters in a text? In fact, digital representations of text require different skills from their reader than analog representations do. This statement will not come as a surprise in the case of visualisations so complex that they require skills in statistical analysis, but it is to some extent true for all types of digital representations of text.

When you have a traditional, analog page in front of you, be it handwritten or printed, you have been taught in which direction to usually read it. You have been trained in school since first grade to do precisely that. For most readers directing their attention to languages they know, it is as though the instructions have been inscribed on the page since

---

41   See Baillot and Lassner, *Von Graphen* [18].

42   See the homepage of the digital edition of August Wilhelm Schlegel's *Correspondence* (`https://august-wilhelm-schlegel.de/briefedigital/`), or looking for instance into Goethes *Faust* on the dedicated page of the DraCor platform: `https://dracor.org/ger/goethe-faust-in-urspruenglicher-gestalt`.

they have been familiarised with deciphering them since their childhood. They know the code. The content will make sense if they read the series of characters in a specific order that was taught to them in school and has been reinforced in their everyday environment ever since.

Things are different when it comes to digital text. There are several forms of digital text representations for which literacy taught in school will not suffice for literate readers to generate meaning. I will examine this more closely, starting with what is maybe the most important step in the production of digital text: the shift from an image format to a text format. This is a process in which machines have been intensely trained in the past decade. Computers have become better and better at recognising letters from images of printed pages (optical character recognition, OCR), or even from manuscripts (handwritten text recognition, HTR). They are trained to identify first what makes up a line, then to duplicate the same line structure in a text format for the output. Within the line blocks, they then isolate groups of characters, which generally form words and, within these units, their next job is to identify, on the one hand, the shape of the characters and match it with existing characters, and on the other hand the combinations of characters that are possible. For instance, if the computer's character recognition system identifies that a unit that looks like a word and is composed of something that looks like four letters, with the first one being an "r "and the last one being a "d", the computer will look up a dictionary where frequent words are listed (ideally, a list tailored to the context of the text concerned, which suggests high probabilities of usage for the vocabulary specific to this text) and indicate a textual transcription that would be, for instance, the word "read", with a high probability of occurrence if the concerned text is a primary school manual in this case.

The process of generating a complete transcription of a text based on an image requires the image to be of extremely high quality and the computer to have been programmed and trained for the specific typography or handwriting that is being used. It also needs tailored dictionaries that facilitate statistical assumptions for word probabilities. Even with this technical apparatus, the output will not look like what is generally considered a text. There will be mistakes, words that are not recognised, diacritics or punctuation marks that will be identified as letters and vice versa. The textual output will need to be improved, in the end, manually by a human, in order to comply with the standards of what is considered a legible text. Generating legible text from a digital source requires special skills, just as deciphering old manuscripts does.

The process of automated text recognition[43] is in many ways exemplary of the necessity of setting up modes of collaboration between humans and computers in order to gain legibility. It is also symptomatic of some misunderstandings that may arise along the way, which is why I will now look more precisely into the process of acquiring text from an image.

The scan of a text page is a visual representation of this text. Unless it has been either programmed to recognise the characters or informed regarding the textual meaning of the image it is displaying, a machine is unable to derive text from the image of a text. This also has consequences at metadata level: if the image of a text is not connected to textual information (be it the wording in textual form or information about this text in the form of metadata), it cannot be found on the internet unless one knows its precise URL. An image without comments is invisible in the digital context. While it might be too tedious to fully transcribe the content of a scanned page, it requires only a little effort to describe it in a few words: "this is an image of a page in this or that script, written by this person in that year, dealing with topic so and so, and which can be found in this place". With these few accompanying words, the image is provided with metadata, and it can be found online.

Now let us assume that the scan of a text can be found online thanks to its metadata. Many readers will say the image suffices as long as they can read it. They will consider that it is legible if the image resolution is of a good enough quality. Poor reprint editions are doing nothing else than taking for granted the fact that a good enough image of a page is a text, a published text even. Readers, deeming themselves happy with the online scan, can then either read the scan cursively online, download it onto their device (in order to read it on an e-reader, for instance), or even print it. In that way, they will have been able to gain access to a text that may not have been within their physical reach, if, for instance, it is preserved only in an archive or a distant library. In terms of access to text, the reader has gained a lot.

---

43   Automatic text recognition, ATR, includes both OCR and HTR.

But the procedure is still problematic in two ways. First, not any random metadata will make it possible to search the web with success for the text the reader is looking for. How is it that by searching for the title of the book, sometimes even an erroneous form of it, you still manage to find it? Because the underlying information is structured — with both positive and negative consequences. A whole economy is based on promoting specific internet resources as a result of user requests. The algorithms used by major companies active in this area have been programmed to recognise, analyse and filter metadata in such a way that users keep seeing their products, with web browsers generating information that will in turn foster the economic value of the information offered. While this primarily concerns other types of (cultural) consumables, even domains like text dissemination have fallen prey to it.

Access to quality text is a particular challenge because quality is not the primary criterion for the companies that decide what it is that your browser-based search will lead you to. Once you have found a text, you cannot easily know whether it is reliable or not. And is this text really what the metadata tells you that it is? Sometimes the result of an internet search appears to have very little to do with what you are actually looking for. The unsorted wealth of textual material that can be found online requires proper training for the reader or user to recognise online text quality, especially considering the current lack of standardised ways to display who the author is, what the title is, in what context the content has been created, and whether it is acceptable to reuse it for whatever purpose. This information can be retrieved in an online library or archive catalogue, but not on every webpage. There are initiatives to make these catalogues more visible, and to direct readers towards them when scrolling the web for information. But making them even more visible would require a massive investment, similar to that which, for instance, the institutionalisation of archives must have necessitated in their time — which was not solely an issue of time, money and space, but also of political will.

I would like to go into more detail about what makes for quality in a digital text. Findability is one of the key criteria. Legibility by a human reader is another, as is its legibility by a computer. If a human reader is able to read the poor result of an OCR output, for instance, they may be able to improve it, and in that manner make it accessible to even more readers. This means that improvability of the text that is made available to readers is another quality criterion.

This is one major difference between born-digital and digitised texts. From the traditional perspective on a work of literature, one would identify a specific iteration as the authoritative one: the one and only version of reference. Compared to this version, there can be only lesser versions or variants. A born-digital text, on the other hand, can very well have several versions. These can be identified with a version marker (in the sense of the master versions I mentioned in section 1.1.2) and a time stamp. The time stamp is not temporally unidirectional: it does not assume that a later version is necessarily better than an earlier one. But it makes it possible to identify versions unambiguously and distinguish versions generated at different moments in time from one another. Applied to the inherent history of the text that I presented in section 1.2.1, it means that it is possible to preserve different layers of the constitution of one text, with metadata providing documentation on the status of these different layers. Digital editions of textual heritage can take advantage of this plasticity to elaborate a more dynamic concept and form of text.

There is yet another quality that is required from a digital text: it should be connectible. The power of hyperlinks lies in their ability to foster discoveries not simply from one website to another, but also from one specific element of a webpage to another digital resource. A website containing many stable hyperlinks is a door to a wealth of other resources. The best-known way to do so in the most common browser language (HTML) is to have a hyperlink underlying a word or a group of words: if you click on the highlighted word or group of words, you are redirected to the resource corresponding to the underlying hyperlink. But this is not by any means the only way to connect two digital resources with one another. There are other types of connections from one text to another text, some requiring the use of other computer languages.

There is more to this question of the digital connection of text-based resources (text-based here meaning that there is always text involved, either at data or at metadata level, or both), and it is worth a closer inspection for several reasons. One of these is that it shows the type of effort that can be undertaken to balance out mechanisms based on the economic value of information exchange and dominated by big tech companies.

Some digital information is not connected via simple hyperlinks, but through an elaborate system of unique identifiers that are being used by a variety of actors of the web in such a manner that they serve as standards, either because they are official standards[44] or because they are widely used and have become *de facto* standards. Cities for instance, can be identified via geo-coordinates. If your great-great-great-grand parent mentions Paris in the old papers you found in a drawer of the family home, how can this be linked to additional digital information? Geo-coordinates can serve as a pivotal information for the machine to connect information from one specific webpage with, for instance, a platform like Wikidata, which centralises basic information.[45] For this information to be recognised by the machine (and interpreted as: "this word is Paris, meaning the city of Paris, France"), it matters to integrate geo-coordinates as underlying information or metadata to the related passage in the text in this case, the word "Paris" in the original text, and to do it in a standardised manner that will be legible to the computer and make it possible to generate a connection to a central platform like Wikidata.

If the old papers in the drawer are not the manuscript of a novel, but something more likely, such as letters or ledgers, the chances are that not only places will be mentioned, but also people: friends, family, and public figures at the time of writing. In the case of persons that are mentioned in a text, the connection to information is even more refined.[46] Librarians all over the world have joined forces in creating a colossal online catalogue in which writers, and to some extent also other personalities in the book market such as publishers and critics, are each assigned a unique identification number. This resource is growing by the day and is less and less limited to book market-related figures, encompassing more and more historical figures in general. The goal of this huge index is for all digital resources to be able to use a unique identification number when referring to a specific historical person, facilitating the connection of available information about this person. Digital resources that use this person identification system, such as libraries or digital editions, can register on

---

44 Like ISO Standards, see `https://www.iso.org/standards.html`.
45 See `https://www.wikidata.org/wiki/Wikidata:Main_Page`.
46 For a more technical description of the following, as well as potential applications, see Baillot and Busch, *Vernetzung* [13].

another Wikipedia-based service called Beacon, which connects all of the registered users through the uniquer identifier, every time they use it — and so on and so forth for every single occurrence of every single person mentioned and tagged with a common unique identifier.[47] In return, each resource can automatically point to the other resources that use the same system. With the Wikipedia Beacon, biographical data can be aggregated while keeping a precise track of their origin.

Catalogues containing biographical information about the people involved in the production of books (writers, publishers, printers, translators, illustrators and critics) were initially curated at a national level. This valuable information stock is now merged into an international digital catalogue.[48] Today, we can resort to a world-wide standardised resource identifying writers — meaning here people that are relevant to cultural heritage institutions and involved in writing processes in one way or another — and more and more other (known and lesser known) historical figures. Of course, things are more complicated in practice than they may seem on the paper. There now are so many data providers that the catalogue contains many doublets. Quality control is more complicated on a world-wide scale than when dealing with a small institution's index of persons. Additionally, more and more non-librarians and non-archivists wish to contribute information, to generate unique identifiers themselves when they gather information about interesting figures of the past, especially scholars working on earlier periods for which it is necessary to process archives in order to gain novel biographical information. Their input is not always as standardised as that of professional librarians and can generate some confusion.

But even as imperfect as its implementation at a global scale may be, the benefit of such a standardised approach cannot be emphasised enough. As a reader, one does not need to engage in intensive research on each person that is mentioned in a given historical context, one simply has to use the identification number and rely on linked resources. The remarkable level of connectivity this entails makes it possible for digital texts to lead from one information source to the next: they draw a path

---

47   See `https://de.wikipedia.org/wiki/Wikipedia:BEACON`. See also Baillot, *Krux* [9] on methodological challenges to the work with biographical data for network analysis in textual studies, and for a less historical approach including a use case for the Wikipedia Beacon, Baillot, *Weisheit der Massen* [7].

48   See `https://viaf.org/`.

through the web for the reader to follow, in the footsteps of a writer or another historical figure they want to discover, even if this person has been dead for centuries. With the original resources linked as entries for each identification number, it is always possible to fact-check any given piece of information by clicking back to its source.

The large-scale realisation of this ideal web of knowledge will only be possible, though, if textual resources are also made available online in such a way that they are accessible for the reader.[49] This is best illustrated by the example of the book scan I mentioned earlier, assuming that it is only available in an image format, accompanied by minimal metadata that connects biographical information on the author to the scan of the document. In such a configuration, readers can bridge the gap between biographical information and text, but they cannot do much with the text itself, apart from reading it from the beginning to end. If there is a raw machine-based transcription underlying the image, readers can search for specific passages based on keywords or phrases, and jump to the parts of the scan that are likely to be of interest to them, without having to read the book full to find the relevant passages. OCRd books can also feature hyperlinks that connect the table of contents with the corresponding chapters, allowing the reader to move swiftly to the passages they are most interested in.

But, based on an image of a text, the possibility of connecting it with other digital resources remains limited. An image contains no information that can be interpreted by the machine in order to generate a connection with another text. To reach a better permeability between textual resources, it is important that underlying machine-readable information is added to them. This information has to be structured in such a way that the machine is able to interpret which part of the text it refers to (the whole text, a paragraph, or a single word), what type of information it provides (for instance external input, semantic information, linguistic structure, details on text genesis), and whether that information should be matched with external resources, as was the case for the geo-coordinates or the referenced persons I mentioned above. This should be done in a computer language compatible with those used in

---

49   In the following, I present the technical aspects of this accessibility issue, while the economical and philosophical dimensions are addressed in section 2.2.2.

other digital resources with which one wants to generate a connection or a link.

The ability for programming languages to communicate and interact with one another can be roughly subsumed under the umbrella concept of interoperability. There are, of course, different ways for digital resources to interoperate, at different levels, and not all interoperability leverages work the same way.[50] What matters here is the general idea that it is possible to conceive digital texts in such a way that the information contained in one can be integrated with another, that they can enrich each other reciprocally. This supposes, of course, that authors authorise the reuse of the text that they have produced. In the current European legal framework, copyright rules favour reuse of older data (by authors that have been dead for at least more or less 70 years, depending on national jurisprudences); living authors are required to state reuse conditions explicitly. But for my purposes I will assume that reuse is possible, and that there is no reason why someone would not want others to use the information either written a long time ago or provided online with a reuse authorisation, in the form of a license.

With these constraints in mind, what can you do with the old papers you found in a drawer in the family home? The local archive does not want to have them because they are not old enough, or relevant enough, or maybe you do not even want your local archive to have them. You think that these old papers are incredibly interesting; for whatever reason you might have, it matters to you that they are preserved and made available to a wider audience. What do you do in order to make this happen on your own? It is entirely possible to do it with everyday equipment and simple technical skills.

Step one, you scan the old papers and add metadata to the scans, even if only a file name and a date. Next, you need a repository where you store your images: it can be a hard drive or a cloud server, for instance. Then you need a way to give access to the information housed on your repository. At this point, you cannot do without some use of computer languages. The easiest way to provide access is to use a simple content management system (like Wordpress), which avoids confronting readers unfamiliar with computer language with the code underlying the editing

---

50   See for instance the subtleties introduced by Syd Bauman as early as 2011 in his seminal
    Balisage paper [23].

process of a webpage. You will have to check regularly on your content management system, though, to make sure that the links remain active through all the various system and browser updates that will come up over time. But this solution has advantages that make the curation effort worth while. In your content management system, you can add hyperlinks to external resources and, in that manner, generate a connection between your resource and other resources. This greatly increases the findability of your digital resource. Now your minimal webpage presents pictures of the documents that you took with your smartphone, a simple description of the documents and links to complementary resources, and it is ready for everyone to access.

What will you have provided then? In a way, a digital archive, albeit not a very sustainable one. You will have made your documents available and, by adding metadata and hyperlinks, you have made it findable. If you add a license authorising reuse, you will have made it reusable, too. What you have not provided if you remain at this level, however, is a digital edition of your resource.

A digital edition would mean that you had supplied a text (and not simply images and metadata). Such a text can be of variable quality: it can be a raw transcription; it can even be an OCR output that is almost illegible to the human eye. But this is a first step, and can be considered a digital edition if it is conceived in such a way that it can be improved and referenced, and that you have explained how it was conceived. Of course, digital scholarly editions are validated through further quality insurance mechanisms.[51] But, by and large, having established a text (again, not simply scanned an image) makes you an editor. This means, conversely, that it is necessary to provide information about the person (and/or algorithm) that generated the text: an edition is always procured by someone or something that has to be named and characterised since it situates the editorial endeavour. One piece of information that is essential to any digital edition is who is its editor: if that information is not available to you when you consult a digital resource, you cannot take for granted that you are dealing with an actual digital edition.

---

51 For Digital Scholarly Editions for instance, a formal catalogue of quality criteria has been developed with the goal of providing the backbone for evaluation processes. An English version is available at `https://www.i-d-e.de/publikationen/weitereschriften/criteria-version-1-1/`.

What happens at cultural heritage institutions when they are in a similar situation, having acquired, for instance, a new manuscript that they would want to make available to a larger audience? They have one key advantage: infrastructure. They have professional scanning material and skilled personnel, maybe also expensive automated text recognition software, server capacities, and a web interface. They can also rely on a set of standardised metadata that are integrated in digital catalogues, making the resource easy to find online. They can provide information on the content of their manuscript and connect it with the VIAF catalogue (that records writers of all times and other historical figures). They also have the opportunity to add an entry in this database and create an identification number if the author is not already registered.

But again, what is then provided is a digital archive, not a digital edition. A digital edition requires one to procure a text, in this case, at the very least, a transcription of the manuscript, and ideally a transcription that gives further information about the text. This information can relate to the way the text is distributed on the manuscript page, whether part of it has been erased and rewritten (if so, how), and whether it contains allusions to another piece of text, be it explicitly or implicitly (and then, if possible, point to it). In fact, in a digital scholarly edition, it is possible to gather as much information as you want about what I called the geography and the history of the text in section 1.2.1.

This is because, in a digital edition, you are not bound by the constraints of the page format. It is possible to integrate a wealth and variety of information in the digital source document of the edition — more than can be technically displayed on one page or on one monitor screen in a way that could be interpreted at a glance by the human eye. How this information is displayed is another question, and to some extent a secondary one. The core of a digital scholarly edition is not what you can see on your monitor; it is the information contained in the annotated text data (the encoded source file).[52] It is all the more difficult to appreciate what a solid digital edition really is when what you get to see online is not primarily the source code, but one way of presenting it.

---

52 The Text Encoding Initiative has been developing guidelines for the sustainable and interoperable annotation of text since the 1980s. It fulfils all the requirements I mention above and facilitates those I mention below. For an introduction to the TEI, see Burnard, *What is the Text Encoding Initiative* [31]. The TEI Guidelines can be consulted and browsed at https://tei-c.org/guidelines/.

This way of presenting the information will inevitably be partial, constrained by the limitations of the monitor interface. But whatever the design choices digital editors make, they have to explain what is presented and how, and that, again, takes up space. As a result, one crucial accomplishment in the design of digital editions consists in finding a balance between, on the one hand, the intuitiveness of an online representation, which is key for the reader to gain orientation in the presented text and, on the other, all the visual options one might have available to represent the complexity of the textual phenomena one wants to account for. This might be interlinear, non-textual information, for instance, or changes in colour and disposition.[53]

To some extent, and as opposed to other devices such as a smartphone or a tablet that can be moved and turned, personal computers (PCs) still make editors largely dependent on the page format when editing primary material that is displayed on a page in an original manuscript or print (the analog source). If you want to edit text that is written perpendicularly to the rest of the page in order to display it on a PC monitor, there is no choice but to turn it around in your online presentation so that it is aligned and in the reading direction relative to the rest of the page. Otherwise, readers would have to turn their necks 90 degrees to read it in a position corresponding to the one it has on the manuscript (which is not a problem with a book, as you can turn a book through all 360 available degrees).

In that sense, when it comes to digital editions of textual heritage, we are only partially emancipated from the page format. In fact, many digital resources reproduce even the gesture of thumbing through pages when presenting scanned books. But this is only a cosmetic issue. A much less cosmetic question concerns the way to cite a digital text. When it comes to citation, the page is most certainly convenient. Or, more accurately, citation practices have historically developed alongside page and book formats, and not staying with such units in the digital world makes text citation much more complicated.

Depending on the device you use and on your personalised settings, an electronic publication will be displayed on a varying number of pages, which means that the page disposition can change according to your

---

53   See Bleier et al., *Digital Scholarly Editions as Interfaces* [25].

personal settings. Page 3, for example, could refer to a variety of contents and not, as is the case with books, to a stable block of text that is printed (or, for that matter, handwritten) on the same page as the number three. In order to refer to the same block of text independently of the device on which it is displayed, it becomes necessary, in a digital context, to rely more strongly on overarching structures such as chapters and on underlying units such as paragraphs. It has therefore become good scholarly practice to quote paragraphs from publications. At first sight, applying this system to, for instance, literary texts, seems a bit dry. But the only alternative at present is to rely on the URL, independently from the length of the text one can find under this URL. The notion of page remains subject to many variations when applied to the web.

The fact is that the emancipation from the page is only partly achieved to this day, and that no alternative practice of constructive scrolling, no serendipitous move from one online text to the next, has yet established itself as a recognised cultural practice.[54] How to read online text is still something of a mystery, also because readers often do not know how to read, given all the features, columns, and banners, and how to know what it is they have read (fake news or valid statement?).

This consideration brings me back to the question of the qualities a digital text should display. We now know this much: it should be findable, interoperable, enrichable, citable, and legible for man and machine. At this point, it is legitimate to wonder bluntly to what extent these quality criteria are compatible with those developed for print editions or traditional archiving practices. A digital text does not need to be completed, it simply needs to state where it stands in the realisation process. A digital text does not need to be authored by one person, but the roles and contributions of all the contributors should be stated. A digital text needs to be curated in terms of the software necessary to access it, otherwise it might cease to be consultable. A digital text cannot subsist in the long run without a stable infrastructure, in which it is different from books but similar to archival material. One of the consequences of these shifts pertains to the constitution and transmission of texts that are recognised as central

---

54   In the article entitled *"Lecture"* (reading) of the philosophical dictionary *L'interprétation*, Denis Thouard assumes an ongoing renewal of reading practices as a fundamental cultural device: "Dans son universalité, la lecture est un mode d'orientation fondamental de l'être humain" [99], p. 259. See also François Moureau, *La plume et le plomb* [83]; and Brigitte Ouvry-Vial's stance on reading seen as commons [84].

by the concerned cultures. In particular, it fosters a reassessment of the literary canon and how it is positioned as a cultural practice.

The orientation provided by the literary canon especially over the past decades and centuries relies on archiving and publishing practices. The examples related to Goethe that I have presented in sections 1.1.1 and 2.1.2 are explicit in that regard: there is a political and cultural coherence in the choice of literary works that are considered essential in a given cultural context, and this coherence can be better monitored through control, at state level, of the infrastructures that are responsible for the way these texts are provided to the public. If the Goethe-Archiv is funded by the state, it will be appraised in educational schemes: there is an expected return on investment. Looking at the current state of digitisation, most of the countries do not invest vastly in large national schemes including targeted investment and infrastructure, while, on the other hand, initiatives like Google Books provide a mass of digitally available and, to some extent, searchable text, which it is impossible to compete with at state level. This raises the question of who should serve as gatekeeper of textual culture at large, and to what extent it is problematic — or not — if it is not a public actor like a state, a region or a city that decides and implements what is worth digitising and what is not, and in what quality.

It has been a major effort of literary studies scholarship to take advantage of digitisation in the past decade to propose a revised view of the literary canon, at a European level at least. Much energy was invested in a form of self-criticism, as well as an analysis of the biases in the traditional literary canon, looking for ways in which these biases could be counterbalanced.[55] By doing so, one inevitably creates a new canon. At the time I am writing this book, digitisation of textual heritage is at its most extensive generated by big tech companies. States, and to some extent the EU, try to sponsor specific branches of cultural heritage by funding the digitisation of specific collections. Scholars strive to analyse and channel efforts towards a better-balanced representation of linguistic and/or stylistic minorities, as well as politically and culturally less dominant figures.

---

55 A major impulse was given by Moretti with *Distant Reading* [82], as well as Jockers, *Macroanalysis* [71]. For more recent progress, see the work on the "European Literary Text Collection" ELTeC, `https://www.distant-reading.net/eltec/`.

It is difficult to predict which of these influences will prevail in the decades to come. The tension is also one between the quantity and quality of digitised information. The current efforts speak for a better recognition (and establishment) of texts that are preserved in a more sustainable manner: texts that are more findable, more enrichable, and more shareable.

In that regard, the way in which human readers adapt to reading more than just linear sequences of pages may play quite a decisive role. As important as the cultural technique of linear reading of a book page may be,[56] media studies have shown that other forms of reading have gained traction with the popularisation of web-based content dissemination and, with them, other perspectives on textuality. In the next section, I would like to show how Open Access has the potential to be instrumental in shaping new quality criteria for textuality in the context of wide dissemination and enriched forms of reading.

2.2.2 Access made Open

In the previous sections, I have elaborated on the quality of digital text at what is a rather theoretical (albeit occasionally technical) level, without taking into account the economic trade-offs that digital access to text relies on. I will now turn to discussing access to text taking account of that point of view. This includes an analysis of the way in which digital archives, digital editions, and digital publications in general are interconnected, which provides an opportunity to present all that an open access to textual heritage would ideally entail. In fact, I intend to show all the good that digital media have done to the access of text. I begin by paying closer attention to the philosophy of knowledge sharing, but I will ultimately circle back to technical feasibility.

The modalities of consultation of web content that I have previously mentioned were subsumed under an assumption that I have left unexplained until now. All digital connection processes, whether connecting via hyperlinks, enabling web browsing with the support of geo-coordinates, or providing a wealth of biographical data, only work if the resources the links point to are accessible, that is if they do not require

---

56   On the negative effects of digitisation on bibliophily, see Piper, *Book was There* [89]; and Carrière and Eco, *N'espérez pas vous débarrasser des livres* [35].

the user to pay in order to access them. Many resources, however, are not available for free, but hidden behind a paywall.[57]

To some extent, it makes sense to have to pay to access content. You pay to purchase a book or a newspaper, and you sometimes pay to visit a museum: the personnel and infrastructure that provide content and service are working to make it possible, and work has to be paid for. But you do not always pay to use a library (or only very little), and you seldom pay to consult documents in archives. This means that mechanisms giving free, or almost free access exist, even to material that is produced in a private context and involved funding in the production of information. And there are certainly mechanisms to grant free access to material that is produced in a public context. Archives curate documents produced by administrations whose work is paid for with taxpayers' money. In return, taxpayers are provided with the opportunity to consult them.

Over the last centuries, such trade-offs (for information that is generated by public servants) or forms of leverage (for information produced by free market actors) were developed to foster access to information. They related to material of a textual nature in the case of shared resources that are considered of general political and/or cultural interest. Political interests then translate into economical means to facilitate access: for example, states subsidise cultural heritage institutions so that these can share their material. Not all heritage institutions share material under the same conditions: the balance between public service and economic sustainability is not easy to maintain when one has to take care of a building, employ personnel, pursue an acquisition policy, curate collections, and enrich catalogues. Investment choices depend on political agendas, but generally, it is considered one of the tasks of the public administration to monitor and facilitate access to relevant textual heritage material, whether it is in the form of archives, libraries, or museums.

What is true in the analog world is also true in the digital context. But since the digital has not been available for as long a time as public archives, libraries, and museums, it still lacks, at least to some extent, safeguards that would make the question of access one that is easy to answer. Access to text involves a series of actors and mechanisms whose

---

57   To my great dismay, this is the case with many of the resources I am pointing to in the bibliography of this book. I would have liked to be able to provide only fully accessible references, but it would have required the sacrifice of many important ones.

interests and whose business models are likely to follow different rationales. As a consequence, there are a variety of ways to compensate for the costs corresponding to the effort of making text available online. These generally depend on the type of text and on the type of institution that provides it.

If public archives wish to make their material accessible, their chief goal has to be to ensure sustainable access to metadata, in such a way that they can be integrated into a meta-catalogue, for instance, a national finding aid of all literary archives such as Germany has,[58] and in the best case to images of their textual material also. If libraries want to do the same, the requirements are similar: an online catalogue and an online collection of book scans make it possible for readers to find and read the books they are looking for.[59]

None of this relates to an edited text in the sense of a digital edition. In order to make a digital edition available, one needs to give access to full text in a text format and not simply as an image, and to acknowledge editorial choices explicitly: why this version rather than another, why information is displayed the way it is and, ideally, with what other online resources the digital text is connected. The digital infrastructure as well as the human skills necessary to produce an edition are different from those needed for a digital archive or library. But they are not necessarily disconnected from one another. Indeed, they are complementary and should be connected. When you find a digitised book, you might want to be able to quote passages without having to transcribe the whole book for yourself, and you might want to know why this version has been digitised rather than another, and also, perhaps, understand why this library has this book, in the case of a rather rare volume.

This is all the more true for archives. Finding aids there usually contain such information as type of document, number of pages, date, place, writer(s) and acquisition history. Ideally, this information is detailed for every single item, but even such a minimal set of intelligence requires quite some effort to assemble. Generally, archives receive archival material in the form of a folder or a box, in which papers are in a specific order

---

58 See `https://kalliope-verbund.info/en/index.html`, a unique centralised resource in the German context where federated structures usually lead to a wider distribution of resources.

59 See, for instance, the French Gallica website: `https://gallica.bnf.fr/accueil/en/content/accueil-en?mode=desktop`.

that is not necessarily the most logical one (letters, for example, might be thrown together without regard of their chronological order). Following the provenance principle mentioned in 1.1.1, archivists need to record the way in which the folder was handed to them, and number its elements, then consider the form in which the folder has been acquired and align it with conservatory or archival logic, and then record this information. Even if only to proceed to this work step, they need to carefully consider each textual unit and its relationship to the whole folder. Recording information for each single document adds one more work step. And making them available not simply through the in-house catalogue, but for a wider audience via catalogues that are freely accessible online, for instance, requires yet further effort, and a complex one too, since it is necessary to coordinate with other archives in order to align finding aid systems. [60] Needless to say, not all archives can afford all of these work steps.

When you look for a precise piece of archive, you rely heavily on the quality of the recording achieved by archivists. If you do not know which institution is likely to host that piece of archive, it will be crucial that their in-house finding aid is integrated in meta-catalogues that regroup finding aids from archives with similar interests, otherwise you will first have to list the relevant archives, then call or e-mail them, or even visit them, in order to see if the manuscript you are looking for is actually there. When I first started working on 19th-century manuscripts, state of the art technology was no more than a central Berlin address to which you could write a request letter and who would photocopy and send you home the index cards of the manuscripts you were looking for. Their major performance rested on having copies of index cards from all over Germany. We lived in a world of index card drawers. To make sense of them, one needed to know what item one should look up in the first place. There was no such thing as a plain text search: the user had to be familiar with entry points and bibliographical conventions.

I received a list of the manuscripts relevant for my research around the year 2000. I then visited archives, touring north and east Germany, on my manuscript quest. For some documents, it was not really worth the effort of traveling to consult them, when only a couple of pages proved to

---

60   Obviously, coordinating at an international level is even more challenging.

be relevant to my research. For those, I acquired black and white copies that were sent to my home by the archives for a moderate fee. In general, this sufficed for the quality of transcription I was aiming at. I mostly simply wanted to transcribe, not procure a full-fleshed edition. But from the moment on that you start making decisions in the way you render the characters you can see on the manuscript page, you transform into an editor. Should I expand abbreviations? Indicate who I think it is when names pop up? Explain what date is meant in the mysterious doodle on the corner of the page?

It did not occur to me at the time that it would be possible for my transcription to be easily placed in relation to the index cards in the grandiose drawer room in Berlin. What I was chiefly interested in was to find a way to present the comparison between the text I had found in the manuscripts and the edition that was published in the early 19th century, in which passages were missing, names were omitted, and whole parts of the document folder were ignored. I was all the more eager to make my editorial work known to a wider audience because I could see evidence that the person who had published the altered edition had had the exact same manuscripts in their hands as I did myself. They left editorial marks, strike-throughs, comments in the margin, most of them corresponding to the editorial choices made in the print edition.[61] The only things I was really missing in order to fully understand how this edition had been conceived were the proofs of the edition and the correspondence between editor and publisher. I have not found them to this day, but I have come to realise, also, that the work I do has to be conducted in the context of a structural lack of information. I will never be able to fully understand it all. But I am able to understand enough, from what I can marshal, to improve knowledge on a variety of topics, such as intellectual networks, editorial processes, and correspondence rhetorics, for instance.[62]

But then what was it worth to advance this knowledge for the handful of scholars interested in Romantic studies who would have read the scholarly articles I wrote? Why would it not be possible for people interested

---

61 Anna Busch shows in *Visualisierung* [33] how this contrasting information can be extracted from the different iterations of the text and digitally displayed. See also a comparison of different visualisation tools for textual variants by Torsten Roeder in the *RIDE* journal [93].

62 See for instance *Berliner Intellektuelle als Programm* [12], *Netzwerke des Wissens* [4], *Das Netzwerk als Kunstwerk* [8], *Wissen, Lieben* [6].

in the transcription to get to see the manuscript and understand all the different layers of corrections, and to compare editions? And why should archives not be able to provide the transcription I made that could surely be of assistance to other scholars, perhaps in such a way that it becomes interesting not only for Romantic studies scholars? While these were speculative remarks I made to myself — fantasies, really, twenty years ago — making that big dream possible has been greatly facilitated by the popularisation of digitisation processes developed from the 2010s on. If a scan is online, and a finding aid is online, and a library catalogue is online, and a manuscript transcription is online, all relating to the exact same manuscript, then why not connect them all together? What was materially impossible with index cards in drawers, with manuscripts hidden in boxes in dark, cold rooms that could only be consulted indi-vidually, and with editions that were expensive books that did not even bring much reward in terms of academic reputation, suddenly became an evidence in the digital context. Once you have admitted that all of these — the scan, the catalogue entry, the transcription — are iterations of the representation of a text, and that what you want is to make visible all the knowledge that they entail when brought together, in order for a wider audience to be able to see and read that text anew, then it becomes very simple to define and reach a goal — the idealistic goal of providing access. And providing access means contributing to a massive background in-frastructure that supports the circulation of complex information in such a way that the reader can find, display, and use it as easily as possible.[63]

In a sense, the digital media provide the platform to fulfil the promises of Enlightenment, bringing knowledge to a vast array of readers. The crux is that it requires more than simple alphabetical literacy to be able to process online content, and it does not suffice to make quality information accessible online for it to automatically increase knowledge, rationality, and the state of the world. It requires informed digitally literate users. Nonetheless, technical solutions exist that make it possible to connect information and provide access to different representations of text.

This does not mean that we now have all knowledge of things just because we can gather it. It has more to do with arranging the pieces

---

63 On why an archiving and publishing ecosystem encompassing research, cultural heritage institutions, and research infrastructures is necessary, see Anderson, *What are Research Infrastructures* [2] and Borgman, *Scholarship in the Digital Age* [28].

of a puzzle than actually completing it. What is more, the different pieces of the puzzle do not necessarily fit all that well. Some are broken; some do not adjust precisely. The formats and languages used to present information vary from one area to another. Archives have different digital standards than do editors.

One major challenge consists in making archival metadata and scholarly metadata interact with one another. While both are eager to know who wrote a letter and when, archivists and scholars have different ways of making this information readable by the computer and of displaying it online for a human reader. They have different quality control mechanisms (what, for example, if someone made a mistake typing a date: how do you verify and correct that?). They also have different ways of adding new information as it becomes available. Delayed synchronisation results in a great loss: it would be desirable for metadata to be actualised at the same time in the different resources presenting one specific textual resource. If a scholar realises that the text was not written in 1837, but in 1836, because the writer made a mistake, for example, they will know that because of information contained either in the text or outside of it, but it is information that is not necessarily processed and validated by archives. In that sense, while it may be very simple to add a hyperlink or a pointer from one digital resource to another as a way to connect different iterations of the same text resource, the fact that web content is bound to evolve and change in terms of its content makes the connection technically challenging. And it obviously does not come down to finding a technical solution alone: human decisions are involved — in this case, philological ones — in order for the connection process to be fully acknowledged and recognised by all actors. Generally, this is about changing our understanding of truth (additional information can change our knowledge of facts) and the mechanisms through which we validate the way in which we advance knowledge (not all insights scholars may have about texts may be considered as valid).

All the connecting, validating, and presenting of work steps involve time and effort, which someone, in some way, has to pay for. Archival finding aids are paid for by archives, that is, in general, taxpayers' money. This goes for library catalogues too. Scanning their respective holdings already opens a Pandora's box, as it has only recently become part of their missions and generates important costs. Some institutions partnered with

Google in the context of the Google Books project to foster the digitisation
of their stocks. As for editions, they too can also be funded by taxpayers'
money, but generally not on the basis of a long-lasting mission: scholarly
funding is still aligned on print productions that seal the achievement of
a work process — a tangible end product. In other words, digital editions
conceived in the context of so-called "projects" have no way of funding
the long-term digital infrastructure that is necessary for them to remain
accessible in a few years' time. On the one hand, we have institutions
whose core is funded primarily for missions others than digital access;
on the other hand, we have knowledgeable editors who do not have the
required infrastructure to make the result of their efforts accessible in
the medium and long term. And in the middle, we have a reader who
does not necessarily have the skills required to get orientation and find
their way around the overall mass of text that is the internet. So let us
ask again: how can a reader recognise a good text from a poor one, and
how can we foster access to the former and/or discourage readers from
turning to the latter?

A good digital text is one that provides enough information to be
citable: it provides a stable URL or a unique DOI,[64] an author, a date,
and a title stating the nature of the resource as a bare minimum. But
if the rest of the information is only available upon payment, the value
of referencing these core elements is limited. The type of payment required to access a textual resource varies considerably. Some archives,
for instance, established the rule that the first user to request the scan of
a manuscript would pay for the digitisation process, which is not always
very expensive unless you want to scan large amounts, and the archives
would then add the realised scan to their digital collection for anyone to
access for free. This has two positive consequences. First, it makes the
reuse of existing scans all the more appealing, as users do not have to pay
for scans that are already available: it fosters the reuse of digitally available data. Second, making these texts available in high quality and with
quality insurance as part of a collection hosted by a reliable institution
makes them more visible. The users who paid for digitisation will have

---

64   DOIs are based on a registered ISO standard; see `https://www.doi.org/`. Current
      dissemination practices lead to the attribution by publication aggregators of several
      DOIs to the same text entity, which is extremely detrimental for archiving, cataloguing,
      and dissemination.

contributed to the integration of this textual content into robust digitally accessible resources. In that sense, they took an active part in shifting the textual canon towards what is accessible freely and of good quality.

In other cases, you have to pay either for a single visit to an online resource, or for long-term access. Some offer the opportunity to grant access at an institutional level, in general for a rather high price and as a package in combination with other resources. Universities can thus acquire sufficient entry points to some resources that all their personnel and students can access them, as in the case of scientific journals. In that scenario, individual readers or institutions pay for the service that is provided to them, and thus for the maintenance of the publishing infrastructure. This is a situation very similar to that of the *Ancien Régime* where only a few fortunate rich or well-connected people could have access to cultural resources. The centuries that separate us from that period are precisely those in which public institutions like archives and libraries have been established in order to disseminate access to knowledge and culture more widely. Just as it has become possible to organise sustainable solutions for public libraries that provide access to valuable books, and for archives that provide access to manuscripts, it should be possible to have a sustainable digital model for the public accessibility of textual material even if its production comes at a cost. These models are the ones fostered by Open Access.[65]

Open Access strives to make information available and accessible online for any user. Some forms of Open Access are not as open as they hope to appear in the sense that they simply shift the weight of the cost from the user's shoulders to the producer's.[66] It is the person who wants to display content online that has to pay for the access to be made open and free for other users. Drawing a comparison with the classical library, this would mean that it would not be the library that pays the publisher for the purchase of a book, but the writer who has to pay the publisher for them to deliver the book to the library.

The shifts involved in the relationships between actors in the field of text production affects the mechanisms that frame the collaboration

---

65   On Open Access in general, its different models and their implementation, see the standard work by Peter Suber, *Open Access* [98]; and more recently Avery et al., Special Issue on OA of the *Journal of Scholarly Publishing* [3].

66   On the different models especially in the Humanities, see Eve, *Open Access and the Humanities* [47].

between author and publisher. The added value of a publisher's work — producing a valuable print artefact — is shaken by the potential dissemination power that digital media gives back to writers. Authors can master editing tools and produce legible texts on their own with their personal computers. In terms of dissemination, new media make it possible to catch the attention of readers without relying on a publisher's contacts and advertising. Moreover, readers' expectations in terms of how professional textual output is, as regards typesetting, fonts, and layout, can be lower than the standards expected by book lovers. All in all, you can independently write your text, edit it on a blog, advertise it on twitter, and get a fantastic readership. Monetising this, of course, calls for additional skills. But, by and large, there are ways for an individual to achieve even this to a certain extent, without the help of a publisher.

Even in a context where anyone can publish anything, and perhaps even more so in that context, there remain differences in the types of texts that are published independently, certainly when we consider literary texts. Cultural heritage narratives tell another story than that of a blogger writing fan fiction. And this brings me back to the question of the canon.

Different mechanisms contribute to the inclusion of a text in a literary canon. In referring to texts becoming part of a canon, I mean texts that are to be made available to a large audience based on their relevance for culture and society at large. Obviously, textual quality and overall message play a role, but publishing and archiving strategies do too. Digitisation has become another factor in this process, entailing shifts on account of the opportunity it offers to make extremely large amounts of text accessible to extremely large amounts of readers. Not all of those who can technically access digital text collections will have the skills required to read the digitised text, however. It will be of little to no use for a reader to access a scan of a text written in a language they do not know, or in a script they cannot read. This is an extremely limiting factor if you consider that, even for German texts of the 19th century, for instance, there are fewer and fewer readers proficient and trained in reading the Old German script that was used at the time. How much more difficult it is to make sense of texts from other cultural areas and earlier periods then!

One of the interests of making a text version (and not simply an image version) of such texts available is that they can be annotated in any language with semantic explanations or even with a translation. *Linguae francae* such as English or, in some parts of the world, French or Spanish, can bridge the gap to help lower-resourced cultures gain traction and visibility. Admittedly, relying on colonial infrastructures is not likely to shift the canon very much.

But for such a wide recognition of cultural material, it requires that it be accessible and free around the world, be it only because money has a different value in different places around the globe. The few dollars that an affluent scholar can easily afford in the USA are an excruciatingly high price to pay for someone living in the Global South. So, what would it take for the dream of an open access to cultural heritage to come true?

An internet of cultural heritage should, ideally, be a meta-archive and a meta-library in one, and facilitate the way in which users navigate from one text to another. It should, especially at metadata level, add explanations in languages that are used for global communication: even if these languages are markers of a colonial past, they are often our only tool to ease communication and transmission across cultures. It should also generalise the use of standards in order for these resources to be easily combinable. Ideally, when you find old papers in your family home, you would then take a picture of them with your phone, add the information you have (date, number of pages, and what you can gather from the writer and the purpose in writing), and place it online in such a way that an archive would be able to provide a link to it if it complemented its own holdings. Of course, the photo you took with your cell phone will not be as good as the high-resolution scans archives can display and with which your material can be connected. But, in terms of identifying where sources of information are, and whether it might actually be worth starting a digitisation campaign for your manuscripts, it still helps. One of the big shifts that has taken place over the past years, in terms of digital resources that cultural heritage institutions host and curate, is that they now sometimes engage in preserving and presenting scans of material that does not physically belong to them, but that is a valuable complement to their own stock. Digital archives and libraries are institutions, but they are also infrastructures, and they can connect material that goes way beyond images of what they possess under their own roofs.

Again, this circles back to the question of the strategic choices that are made. Which text should be valued and integrated into digital libraries? Who pays for this? How do we make it sustainable in the long term, and what about choices one comes to regret later? Maintaining a digital collection comes at a high cost and can be subject to re-evaluation. It may well be that, at some point, a library decides that a specific digital corpus is not worth keeping on their server, but should make space for a more relevant one. There is always room for shifts in what is made accessible and what is not, and hence room for the disappearance of heritage material, even though it would be technically possible to preserve it. Digital media do not alleviate the dialectics of preservation and destruction that is inherent in the curation of heritage material, even for a basically two-dimensional item like text that can readily be handled in formats that are easy to store, easy to connect, and easy to share.

If one compares, on the one hand, a digital text published by a publisher, made available only if the user is willing to pay money (that is, hidden behind a paywall), and, on the other hand, a digital archive or library collection providing free access to scans and metadata, the second is the most sustainable one, because it relies on shared, public infrastructure that itself follows standards and is conceived to address a long-term preservation mission. In terms of the depth of the textual information, it is very likely that the published text behind the paywall provides more insights than the simple catalogue with the scans. But there is no way to easily connect the publication behind the paywall to other resources, whereas it is possible to provide free access to digital editions that are built on top of, or at least linked to, a digital library or digital archive. The interaction between these different points of access to heritage textual material is facilitated by the use of common infrastructures and digital service providers. This means that all the technical decisions that are taken at this level are much more than technical decisions: in the middle and long term, they impact what will become the digital canon of reference, and hence what we teach, and what we know.

A digital text of high quality is one that is freely available to all, in a format that makes adding further information possible, in the form of annotations, of hyperlinks, of transcriptions, of visualisations, of sonifications, and more, and which can be submitted to quality control, such as scholarly quality assessment. In order for this reciprocal enrichment of information to work, another condition has to be fulfilled: reusability must be authorised. You will remember how annoyed Goethe was when publishers reproduced his works without him being informed about it.[67] Copyright has made some progress since then, but without evolving very much in the digital context. Authors who are still alive might want to provide reuse rights, but sometimes they are not able to do so because they have, per contract, handed over the rights to their work to a publisher. Works of authors that are dead enter public domain some 70 years after their death, and can then generally be reused without any restriction.

But this concerns the text, and not the single copy or the physical media form that supports it. For the single copy of a book or a manuscript, the person who possesses it also has something to say. It can very well be that your cousin, who owns the piece of furniture in which you found the old family papers, does not agree that you can take a picture of them and put them online, for whatever reasons. The artefact itself belongs to someone. Similarly, it is very likely that the publications behind a paywall are not conceived to be reused, but protected by restrictive rights limited to the publishing house.

While some textual material is undoubtedly sensitive, for personal, historical, or any other reason, and it is also true that not everything should be made easily available, it often happens that access to text is restricted for reasons that have nothing to do with the security or integrity of individuals. In western civilisations, the fear of being deprived of one's creation, and robbed of its value, often trumps the joy of sharing it, probably as a side effect of decades of fierce capitalism. That is not only sad; it is problematic.

If we could assume that the basic conditions for the reuse of a text or of the image of a single copy are that the author has to be cited when referring to it, things would be simple. But in fact, most of the time, when you click through a website, it contains no information whatsoever about the conditions of reuse of its content. Can you make a screenshot? How long a passage can you quote from it? Who is the author? Can you harvest and process the data?[68] If you cannot do any of this, even a text that is presented in a text format is not of much more use than a scan of a text in an image format.

---

67   See section 2.1.2.
68   See Baillot et al., *Publishing an OCR* [17].

This was the last element to add to our assessment of what quality digital text is. A quality digital text is found online in a processable format; it is easy to find through meta-catalogues, and easy to access for free; it relies on stable infrastructures, uses standardised computer languages, and can be enriched with additional information; and it informs the reader about its reuse conditions. Digital texts follow quality criteria that go beyond and add up to the editorial norms that were developed over the past centuries. Quality digital texts are sophisticated, and they reflect the construction mechanisms of society at large. I cannot simply write that this is what the future will judge us on because there is so much more to our culture and society. Yet undoubtedly one of the things on which future generations will be entitled to judge us on is our ability to preserve and transmit quality digital text.

In an analog context, access to text is provided to a much larger array of people through publication than it is through archiving. Textual decisions are taken at each step of the writing, preserving, and distributing of a text, involving a range of techniques and of actors. In the digital context, the limits that separate the different missions in this editing and disseminating process fray. A new division of roles, of functions, and of the money invested, but also of the goals pursued at a political level, could give hope for a future in which high quality text, defined along the lines of sophisticated technical and editorial criteria, could be available to anyone.[69]

The historical perspective in this chapter has shown how a wealth of actors have always been involved in shaping textual products. These collaborative processes have long been made invisible by authority concepts that have dominated the European book market. In the 20th century, media shifts have paved the way for an acknowledgment of iterative and distributed approaches in text production, constitution, preservation, and dissemination. While educational schemes have not kept up with technical progress, we can now marshal all necessary leverages to embrace digital philology at large and champion the cultural, political, and social benefits of access to text.[70] Recent perspectives on the literary canon

---

69   See also Adema, *Living Books* [1].
70   See also the image of the leaf falling on the river in the introduction to the excellent *Digital Technology and the Practice of Humanities Research* by Jennifer Edmond [46].

demonstrate an authentic effort of critical self-reflection which bodes well for future generations.

In the next chapter, I will explore the limits of this — admittedly somewhat optimistic — point of view in the context of the climate crisis.

3. What the climate crisis does to text

Several shifts in access to text are made possible by digital technologies. Many of them have the potential to improve curation and distribution dramatically. They make large quantities of text available, including text that is relevant for cultural heritage. Monitoring choices made by major providers of digitised text worldwide is still an issue in several regards, among others in terms of technical choices, such as formats, ethical premises, and selection mechanisms. Nevertheless, quality criteria are being developed, and are more widely acknowledged by the day.[1] Quality insurance remains a challenge, but with the development of communities of practice dedicated to improving it, it has become possible to work towards constructing digital resources that will be accessible to anyone from anywhere in a good text quality. And perhaps more importantly, if the goal is to grant access to text in general, digital tools offer the possibility of considering material objects and their digital surrogates as a whole, and to empower a variety of actors to take part in the curation of them at different levels, connecting analog and digital worlds. Digital information can be conceived, organised, and modelled in such a way as to circulate between institutions: there is a much greater permeability in the process of information curation addressing different versions or representations[2] of the same digital and/or physical document.

Admittedly, understanding how information circulation operates is not as intuitive as simply clicking around on the internet. And it remains a major challenge to include digital complexity in training schemes. Current educational structures are in most cases overwhelmed by the implementation of the pedagogical material and settings that are necessary

---

1   Funders are playing a key role in leveraging towards empowering Open Science practices. Mandatory requirements condition grant approvals at European or national level in Europe. This is often criticised, in particular by social science and humanities scholars (these reproaches are evoked in Baillot and Giovacchini *TEI Models* [16]), but applying a minimal set of good practices inevitably requires the surrender of some privileges.

2   See 2.2.1.

        https://doi.org/10.11647/OBP.0355.03

to achieve a degree of data literacy to enable a wide array of people to grasp the potentials of what is open to them.

This scale of quantity, quality, circulation, and complexity comes at a social cost: not everyone is able to gain orientation in the digital world. It comes at an economic cost: major companies dominate the market, which also means, considering how they achieve their grip on society, that it comes at a cost for our private lives too. These are all aspects worth looking into, which have been addressed critically over the past years — yet stayed unresolved overall.

Until very recently, little attention was paid to the fact that they also come at a considerable environmental cost. In this final chapter, I want to engage in a reflection on this specific dimension and provide some orientation on the environmental footprint of access to both physical and digital text. My goal is not to negate the idea of a free and open access to knowledge and culture for what north-western societies would like to consider as "the masses", simply because it does not fulfil all of its democratic promises. The following considerations strive to envision tomorrow's preservation, recording, and dissemination strategies for textual content in a context of greater respect for the limited natural resources that are at our disposal. It tempers the ideal of universal access, but by no means intends to terminate it.

My argument for the greater respect of environmental issues is not a speculative one. On the contrary, it is anchored in the materiality of textuality. It could be objected that the digitisation of cultural heritage is, in the context of the big picture of the climate crisis, as good as irrelevant. The carbon footprint involved by the digitisation of cultural items has never been considered a key area to tackle, and it is clear that we are not going to save mankind solely by optimising the way in which access is provided to textual heritage. But I would like to show how access to text is strongly rooted in an overall system that can be transformed for the better. Technical solutions and the intellectual grasp of the mechanisms at work can be used to transform the changes brought about by digital opportunities into less damaging ones.

In the first section, I present general issues and challenges related to the environmental impact of access to text; in the second, I elaborate on a concrete example, focusing on the environmental footprint of the book you are currently reading, composed with the help of the publisher, Open

Book Publishers. The self-reflective process initiated in the final section is conceived as a tribute to an editorial tradition that paved the way for a dialogue on convergences of interests between actors concerned with the transmission of culture and knowledge at large.

3.1   The environmental cost of access to text

There are several ways to envision, and ultimately measure, the ecological harm generated by human activity. Greenhouse gases — mainly carbon emissions — are generally in the main focus when it comes to measuring negative impact. It is a valid indicator in the sense that it gives a compelling idea of the magnitude of destruction we have to deal with.

Emissions are a slightly different calculation than footprint. Footprint includes not only local emissions, but also the greenhouse gas output that is produced by the imported goods in use in a region.[3] In a western country, annual per capita greenhouse gas emissions are evaluated at around 5T as I write,[4] while their footprint is 10T when all the products imported from other countries that are consumed or used in the country concerned are included. Both emissions and footprint concern only greenhouse gas, but there is much more than carbon to take into account when it comes to assessing overall environmental impact. But it is more complicated to measure impact on biodiversity or on water resources. General discourse usually focuses on greenhouse gas emissions.

The environmental cost I shed light on in the following pages strives to encompass all dimensions of human impact on natural resources, even if they are not precisely measurable. My goal is to frame the question of access to text and of archiving textual traces in an epistemological context so that it can be redefined, based on the premise that we do not dispose of infinitely available resources. In that sense, I am moving back to theoretical approaches developed in chapter 1, in which I followed

---

3   The Intergovernmental Panel on Climate Change (IPCC) has set up a Task Force dedicated to National Greenhouse Gas Inventories that regularly publishes reports; see `https://www.ipcc-nggip.iges.or.jp/`. For entities such as organisations or territories, tools have been developed to calculate GHG; see, for instance, the *Bilan Carbone* tool: `https://bilans-ges.ademe.fr/en/accueil/contenu/index/page/calculation_methods/siGras/0`.

4   For France see here: `https://www.worldometers.info/co2-emissions/france-co2-emissions/`. My sources in the following discussion will be focused on France and Europe, where Open Data regulations favour the publication of data and tools.

Derrida's and Ricoeur's arguments that considering what we do not have is essential for us to deal with what we have.

The first section reviews the different forms of access to text I have presented until now and lists the environmentally harmful items they involve. I then move to the ways to improve the *status quo* that can be envisioned in order to maintain a text production, preservation, and consumption activity while reducing its overall footprint. The last section proceeds to an assessment of the emitting factors in the production and use of this book, trying to identify strategies and ways to engage with them more widely.

3.1.1 Assessing the environmental footprint of text

Defining quality digital text, as I have shown in section 2.2.2, is not as intuitive as it may seem, even to those who are used to browsing the internet for textual content. But quality criteria now exist. They make a wide access possible and, with that, a renewal in the approach of text, cultural heritage, and preservation strategies. People from the other end of the planet do not need to travel to archives or libraries anymore to consult a manuscript or a book; they can simply browse a catalogue on the internet, find the link to a scan and consult it. Maybe they can even zoom in on the scan and reach a legibility that might prove to be better than consulting the original manuscript. And maybe they can click, from this scan to, for example, an encyclopaedia, via metadata, and gain additional information. They can run automated text recognition software on the image, supply text and annotation, and gain new browsing options. Everybody can benefit from this.

Or so it seems. Upon closer inspection, this perspective restricts access to cultural heritage in many ways, even when the heritage concerned is simple text and not a complex reconstructed 3D artefact. In order to access, for instance, a reliable digital scholarly edition that provides scans of a manuscript, a critical apparatus commenting on it, and links to further resources, a user will need a good end device, whether personal computer, laptop or tablet, electricity, and bandwidth — all things that only well-resourced countries can provide widely. The dream of giving access to text to everyone is fulfilling Enlightenment ideals so well that it is precisely what it is realising: a liberal *Weltanschauung* agenda for European and North American intellectuals.

The type of access that can be fostered along the lines of what I have been sketching throughout this book is access for the rich. While modern societies have advanced technology to the point that they can convince themselves that it is financially accessible to the masses, I would object that they have not really done so. Although the economic cost has been lowered (occasionally requiring strong political measures) and has become acceptable for a wider array of the population, the environmental cost has risen to the unacceptable for the vast majority, if not for all. Considering environmental cost means trying to shift perspectives from a northern-western point of view to a global one too. Who has access to what exactly, and at what cost?

This question has been haunting me since the moment I realised the potential contradiction entailed in Open Access. I had worked for ten years towards providing a freely accessible, scholarly reliable, reusable digital edition of a variety of manuscripts that, with its choice of texts, strove to address shifts in the literary canon.[5] But providing access to high-resolution scans of manuscripts 24 hours a day, 7 days a week, would not really facilitate much for colleagues from less resourced countries, let alone for a wider audience, because many of the underlying technologies are too complex to be implemented on older computers with a poor internet connection. Not only was my edition not really accessible to these users: it probably contributed to making their lives poorer, since the energy required for high resolution scans, animations, and coloured banners is adding environmental impact for a limited informational benefit. In the bigger picture, it leads or will eventually lead to restrictions on their side — electricity shortages, degradation of infrastructures, and more. The technologies I had been using relied on the idea that it was perfectly sensible to use resources (in some respects, a lot of them) in order to make what I considered a better text available. In a way, my use of digital solutions led me to push the boundaries, perhaps even to ignore to some extent the unavoidable tension of having to make choices, of having to define limits to preservation, of accepting that resources, room, and time are finite.

---

5   See my digital edition *Letters and Texts,* `https://www.berliner-intellektuelle.`
    `eu/?en.`   Older versions can be consulted via the Wayback Machine at
    `https://web.archive.org/web/20220000000000*/https://www.berliner-`
    `intellektuelle.eu/?en.`   The current version can be consulted at `https:`
    `//discholed.huma-num.fr/exist/apps/discholed/index.html?collection=bi.`

Defining a course of action based on this observation is not simple. If you try to delineate more precisely the elements that are energy-intensive and that should hence be either banished or at least reduced to a minimum when it comes to digital access to text, you need to analyse every step in their conception, production, and dissemination. In the following discussion, I will go through this kind of overall assessment by looking into three major types of access to text: archiving, publishing, and digital editing. For each of them, I will list the elements that need to be taken into account to assess their environmental impact based on what is hosted, how it is hosted, and how the hosted material is being accessed. My goal is to shed light on the way these elements are embedded in socio-economic mechanisms at large.

Let us begin with archives. Archives are the oldest institution I have mentioned in this book. They have adapted over time to modern requirements while staying true to their original mission.[6] In today's configuration, archives still need to provide a room that is fit for the preservation of paper documents: a room kept at an even, suitable temperature that protects them from heat, cold, humidity, rodents, and other causes of decay. Ideally, the building would be conceived for that purpose and equipped accordingly. Some of the recent archives (or, for that matter, libraries), have chosen to keep their stock underground:[7] this optimises avoiding light and keeping temperature rather low, although it makes things more complicated as regards to the humidity level. The storage room needs to be equipped with shelves, boxes, and a temperature monitoring and controlling system, equipment that has to be produced and installed. It also needs constant support once installed: at least one person has to supervise the physical conditions for preservation and the machines that monitor them. It can happen that any part of the system (heating, cooling, or control) becomes deficient, and then technical support is necessary. This means having yet another person work on the physical preservation conditions. Additionally, an archive will need recording and consultation infrastructure and personnel: for this it requires a room where people can undertake recording and consultation that will be different from the storage room, since the storage room will most likely be cold and windowless, and its optimal conditions are in any event easier to maintain

---

6   See Pataki-Hundt, *Bestandserhaltung* [85].
7   The most prominent example is certainly the Bodleian Library in Oxford. See Legg, *Underground* [75].

if not aggravated by human presence. All in all, the bare minimum for an archive is two rooms and one skilled and trained person to monitor preservation, recording, and consultation.[8] As recording happens in a digital format nowadays, a basic IT infrastructure is also needed, even for an archive that does not provide documents in a digital format for online consultation. I should add the transportation that takes place when documents are brought to the archive, when staff comes to work, and when users visit to consult documents.

A larger archive will not only have more rooms and more personnel; it will have a much larger IT infrastructure for hosting digital material such as scans of manuscripts. It will very likely also have scanning capacity and server space. This means that a third room will be necessary, one for servers, requiring yet other temperature and humidity conditions, involving skills to be maintained, and additional energy to actually function. What is more, this digital infrastructure is highly likely to be mirrored — duplicated for preservation purposes. The various files will be regularly copied in a datacentre somewhere else in case the actual server stops functioning. This requires yet more resources, this time for maintaining the datacentre and for regularly sending information updates to them.

In terms of environmental cost, each of these elements (personnel, building, transportation, IT infrastructure) has an impact,[9] which depends on the way in which it is being implemented. I have not even accounted for the ecological impact generated by the production of paper and ink in this scenario because the quantities available in an archive are limited and are relatively stable over time. But things are different in that regard when it comes to the second type of access to text, publication.

In the case of the environmental weight of publishing houses, some features are similar to that of an archive: they require storage room, with less strict temperature conditions than for manuscripts, but larger storage rooms to store all printed copies of at least one, more likely several books at the same time, more personnel to monitor the production workflow

---

8   In the UK, the National Archives have drafted specifications for the assessment of the environmental impact of buildings and operations, see `https://www.nationalarchives.gov.uk/archives-sector/advice-and-guidance/running-your-organisation/assessing-environmental-impact/`.

9   At country level, greenhouse gas emissions by sector are presented and actualised by the European Environment Agency here: `https://www.eea.europa.eu/data-and-maps/data/data-viewers/greenhouse-gases-viewer`.

than in an archive, and, as we will see, even more transportation and IT infrastructure. But it requires, additionally, a substantial amount of natural as well as chemical resources for producing paper and ink, and machines for at least printing, binding, and packaging.

Taking textual production from the very beginning, I will consider a contemporary author writing a text. They would probably sketch some of the first ideas and drafts in actual, handwritten notebooks, but then move to a computer-based text. This work step would involve: one computer, ink, and paper, even before the text has left the hands of its creator. Being sent to the publisher, the text will be edited (most likely using a series of software), sent back to the author, to be edited by them. This adds up to more computers for the publisher and copyeditor, possibly more software or virtual storage, and e-mail exchanges. When the final version is drafted, it will require other digital skills in order to conceive and implement layout, involving personnel as well as software, and probably installed on yet another computer. The proofs will then be circulated, needing electricity and bandwidth, as well as the book's cover. Once the proofs are validated, printing can be initiated. For printing, specific machines are necessary, monitored by computers, and more machines and more computers to produce the printed book. The printed books are then bound, wrapped in plastic foils, packaged, and sent to the publisher, who then dispatches the copies to author, bookstores, libraries, and critics, accompanied by promotional material — a description in their catalogue being the bare minimum. This adds transport, more printing, and personnel skilled in advertising.

For an online version of a book, a conversion into digital formats such as HTML or e-pub will necessitate additional skills and software in the production cycle, as well as a fitting device at the reader's end, with the necessary software, electricity for consultation, bandwidth for download, and virtual storage for the ebook. For the physical book, reader endpoints also encompass storage of some sort, even if only a bookshelf in a bedroom. All of these output endpoints, whether e-reader, or shelves in a room, should in the end be taken into account for each reader and/or person who purchases the book or accesses it in another way.

While the overall assessment would come out differently depending on a variety of factors, such as the number of copyedits, the quality and quantity of print, and the type of distribution and of media coverage, you can see that in the case of publishing, carbon expenditure includes buildings, skilled personnel, primary resources (wood and water) for the production of paper in large quantities, industrial equipment, IT equipment (hardware, software, and storage), transport of goods, and a wealth of energy to make it all work. In terms of trade-off, it is not simple to evaluate whether completely giving up on printed versions would be the right solution in the long run: while there would be some saved energy in production and distribution, even more people would have to purchase devices such as e-readers that could display the books. This means having to produce more e-readers, and more access to virtual storage that will be solicited by more people. Rebound effects, which force the development of environmentally costly solutions in order to avoid other environmentally costly solutions are challenging to assess. What is more, the use of binary formats in the field of digital publishing makes the sustainability of strictly online solutions uncertain. Will we be able to read an e-pub produced today in 10 years' time? A look at your bookshelves will tell you that you do not even have to ask yourself that question when it comes to a good old book.

What do things look like when it comes to a digital edition, a digital scholarly edition, for instance, that is conceived to be consulted online only, mainly in a web browser? Considering the challenges in measuring its environmental impact, a digital scholarly edition is a sort of hybrid between archiving and publishing a book. Its business model is closer to an archive if it is funded by a public grant. Although it can of course be funded by a commercial enterprise or a foundation, in my hypothetical experiment, I will consider a public research endeavour. The missions of a digital scholarly edition are close to those I identified for a published book, in the sense that its goal is to disseminate text online in a text format, and not, as archives would do, primarily through metadata, enriched in the best cases by an image of the text. There exist digital scholarly editions carried out by publishing houses and some that are edited by archives, but I will consider the case of a digital scholarly edition realised by a team of scholars, in cooperation with archives providing material, but completely independently from commercial publishers. This example is not fictional at all; it builds on my own editorial practice of the past ten years, and uncovers aspects that the other settings that I have mentioned, either archive or publishing house, did not immediately bring to the light.

Procuring a digital scholarly edition usually requires scholars to apply for grants within the scope of dedicated funding opportunities. In my assessment, I will ignore the energy the applying scholar invests in the application itself, but will start counting at the moment that budget has been granted.[10] From there, budget constraints will dictate a series of technical and scholarly decisions that I will also ignore in the following, though knowing perfectly well that what I may present as a variety of choices is usually pre-determined by the amount of money, time and manpower available in the granted budget.

The production of a digital scholarly edition relies on a team of scholars who usually have complementary skills. I simply brushed over the question of skills above when I mentioned archives and publishing houses, but it is worth looking into questions of personnel training in a little more depth. Trying to assess the environmental impact of skilled personnel would require one to evaluate the cost of their training and include it in the calculation. But things are not that simple. On the one hand, someone who has studied extensively comes at a high societal cost since they received an education over a lengthy period; however, because they studied for a long time, one could assume that they will be more efficient at working once they have completed their studies than someone who has not received as much training. Again, evaluating the environmental cost of professional skills requires one to balance elements that are not easy to compare with one another. In the case of a team of scholars procuring a digital edition, there will probably be a principal investigator who is well-trained and experienced, and alongside them, one or more less trained colleagues who are likely to become experts along the way. Training young scholars costs time and energy up to the point when it starts saving a lot of time and energy. The cursor moves between the two phases more or less quickly depending on the personalities of trainer and trainee.

The scholarly team will work in cooperation with an institution providing primary sources (an archive, a library, maybe even a writer), and with an infrastructure providing server space and other technical settings necessary to access the editorial work. This can be a university datacentre or an overarching infrastructure that provides webspace, the setup for a

---

10   The responsibility of funders (admittedly limited to the question of academic travel) is addressed in Bousema et al., *Critical Role* [29].

database, protocols for exchanging information, etc., at regional, national, or international level. Both the primary source provider and the infrastructure will have at the very least rooms in a building, personnel, and an energy consumption that will be dedicated in part to communicating with and providing services to the editorial team.

Within the research team, communication channels will include e-mail, file storage, videoconferences, actual meetings, work sessions at the office and at home, and maybe at a library or archive (involving different buildings to maintain); team members will each have at least a personal computer and a cellphone, probably an external hard drive as well. For a somewhat more comfortably equipped team, there will be additional monitors, headsets, tablets, keyboards, and a wealth of devices that are likely to come in handy in editorial workflows.[11]

Not all devices of the same type have the same environmental impact. For instance, energetic efficiency can vary from one laptop brand to another, or even between models. It is anything but easy to gather detailed information on the impact of a specific digital device. To assess the overall impact of a device (or, for that matter, of a digital service), it is necessary to consult the related technical report called lifecycle assessment (LCA).[12] A lifecycle assessment examines one device such as a cell phone or a personal computer and assesses its environmental impact, taking into account phases of its production, distribution, use, and end of life.

---

11 In order to understand the relative importance of the different elements involved in digital media such as devices, energy consumption, infrastructure, etc., the MOOC "Environmental impacts of digital technologies" is a good starting point: `https://www.fun-mooc.fr/en/courses/environmental-impacts-of-digital-technologies/`. It comes with additional bibliographic resources, see `https://learninglab.gitlabpages.inria.fr/mooc-impacts-num/mooc-impacts-num-ressources/Partie3/RessourcesComplementaires.html?lang=fr`. See also Marquet et al., *1024* [81]. To give a concrete example that encompasses not only greenhouse gas emissions, but the impact of IT at large, see a report by French Association for Network Regulation ARCEP (`https://www.arcep.fr/uploads/tx_gspublication/etude-numerique-environnement-ademe-arcep-volet02_janv2022.pdf`), which shows that terminals are in most regards the main item in environmental cost of production (Table 103), while datacentres are responsible for the major part of energy consumption in the use phase (Figure 32).

12 The Wikipedia articles dedicated to LCAs are of variable quality at the point when I write this. The English article is not considered as not consolidated enough (`https://en.wikipedia.org/wiki/Life-cycle_assessment`), but the French one meets all quality criteria (`https://fr.wikipedia.org/wiki/Analyse_du_cycle_de_vie`).

The first part concerns the production phase. This means looking at every component of the device, and what they are made of. For instance, they are likely to be composed of rare metals, and the extraction of rare and irreplaceable metals not only lowers the overall limited stock of the resource itself, but also impacts water resources in some cases, or the surrounding biodiversity. In a lifecycle assessment, you will find up to 40 criteria such as the impact on water resources, biodiversity, global rise in temperatures or in sea levels, evaluated for each of the elements the device is composed of. The values are then added, taking all components together in order to give a general evaluation. Since this type of information is not available for each device for each brand, lifecycle assessments work with typical or average known values for similar devices. In other words, it is not fully possible to know which device is better than another similar one because detailed information on the production phase is seldom available.

The production phase usually takes place in countries with limited respect for human rights and not infrequently involves slave-like or child labour. How can one account for that, environmentally and, more broadly, ethically? Stepping into the shoes of a scholar who would do their best to purchase reasonably ethical devices with their public funds, it would be difficult for them to make a case for one specific type of device. It would require an excellent knowledge of highly technical parameters, even before considering the use they will make of it.[13]

Lifecycle assessments of digital devices also evaluate their impact during the phase of use (energy consumption), and the end of life. The lack of satisfactory recycling schemes and the overall growing gluttony of digital devices plead for solutions with the longest warranty and the highest level of repairability.[14] This remains mainly an abstract theoretical stance in the case of an editorial team from the northern hemisphere, because the pollution induced by digital devices that have ceased to function is not likely to be of an immediate inconvenience to a European or

---

13   In France, guidelines are now provided at national level.   See `https://ecoresponsable.numerique.gouv.fr/publications/guide-pratique-achats-numeriques-responsables/`.

14   Despite efforts towards the regulation of WEEE (Waste from Electrical and Electronic Equipment), this remains an underdeveloped leverage towards sustainability at the time I write this. European Union regulations on WEEE can be found here: `https://environment.ec.europa.eu/topics/waste-and-recycling/waste-electrical-and-electronic-equipment-weee_en`.

Northern American scholar: digital junk is disposed of in lower-resourced countries. Biodiversity loss and increased sicknesses due to poor disposal schemes of health-threatening components affect their population, not that of the countries who have used the device while it was working.

Let us assume that the scholarly team wanting to procure a digital scholarly edition has found a way to make an informed and reasonable decision on the digital devices they will purchase with their public funds, that the primary resource provider too will have purchased scanners that are ethically responsible, and that the datacentre they work with is as transparent as possible about the energy they use, and has optimised its facilities to lower temperatures in server rooms, for example, or by using the generated heat for another purpose. Now comes the point where scholars specialising in digital editing can legitimately be asked to make informed decisions. These will concern the data format for the source code, the overall architecture of the database, and visualisation decisions regarding the presentation of the output on a web interface.

In order to make decisions, the editorial team needs to address the environmental impact of production, particularly related to the use of their edition: what happens to a user if they want to access the edition? How much energy will it require from them and from the datacentre that will send the information? How good does the internet connection have to be? The environmental cost of maintaining access is very different depending on the technical setup, whether the web content is actually fully available at all times, or whether it is generated upon user request, based on straightforward scripts that are quickly executed. This is what the editorial environment of TEIPublisher provides: the possibility to generate the requested pages on demand, without having to maintain the whole edition online all the time. It has the additional advantage of relying on economical — sober — and sustainable technologies such as XML-based files, and can be installed on servers of large infrastructures.[15]

After a process of information and reflection that is rather long and complex, the editorial team could have found a way to realise their digital scholarly edition. Once the edition is available, they then have to make

---

15   See `https://teipublisher.com/index.html`. In my own work, I use the TEIPublisher instance deployed on the French research infrastructure Huma-Num (`https://www.huma-num.fr/`). The *Digital Scholarly Editions* platform can be consulted at `https://discholed.huma-num.fr/exist/apps/discholed/index.html`.

their work known and have people actually use it. In other words, they have to write articles about it; to present it at conferences to connect it to other scholarly editions. These are additional, environmentally costly work steps — even the choice of the Open Access option will have some impact.[16] And beyond publication strategies, there remains the question of academic travel. Should scholars attend conferences, travelling, sometimes by plane across oceans, to present their work? It adds another source of pollution to the whole process, and questions yet another traditional academic habit.[17]

In digital scholarly editing, each step of the process deserves to be examined under the lens of its environmental impact, to lead, if not to systematic reassessment of priorities, at least to raise awareness of the global impact of the process undertaken in order to give what editors think is the best access to the best text for the most people.

From the point of view of those who make text available, the contradictions that these processes involve can hardly be addressed in a satisfactory manner. On the one hand, for all the actors I have mentioned, be they archivists, publishers, or scholars, the standard *modus vivendi* in north-western countries is that of a fierce competitiveness, leading to an inflation of activity, of production, and of the general visibility necessary for professional survival. On the other hand, a game with as yet unwritten rules that takes account of the global environment tends to go in the opposite direction and requires us to look into things we do not know about precisely; to take time, to minimise efforts, to avoid all things shiny. It affects nothing less than species and planetary survival. The tension between these two opposite aspirations is an unbalanced one. Decades of professional habits have left their mark on the first one, while the other seems to contradict even the idea that there is room for individual leverage on infrastructural questions in a professional context.

---

16   I definitely take for granted that green Open Access, with no barriers and little editorial added value, is more environmentally friendly than gold Open Access, which can only be accessed through a paywall with data tracking, relying on tailored hosting solutions and in-house formats. Depending on their technical setup, diamond Open Access options might be closer to green or to gold in terms of their environmental impact.

17   The recommendations published by the *Berlin-Brandenburgische Akademie der Wissenschaften* and the *Junge Akademie* in July 2022 provide an excellent overview of state-of-the art research on academic travel, together with action points suggestions. See Gerhards et al., *Klimaschutz* [59].

For a random individual in this socio-economic ecosystem, there is no obvious reason to take the long road to sustainability, as it is not paved with incentives, recognition, or better work conditions. Temporalities play against one another.

From the point of view of those who want to access text, much of what will be within their (digital) reach depends on where they live on the globe. Despite what some may want to believe, and despite the efforts deployed over the past decades to popularise access to text (including digital access), it remains a luxury and a cultural marker. Especially the technologisation it relies on is likely to increase a legitimate sense of global injustice. For many, the natural losses such activity causes are more visible than digital benefits such as access to cultural heritage.

This imbalance should invite especially actors, particularly those from well-resourced countries, to revisit the notion of what is "technically possible" in the light of climate justice. I, for one, am convinced that taking what might seem a step back is, in fact, a major leap forward. In the following section, I want to draft a few perspectives on this.

3.1.2 Archiving text for tomorrow

A world in which digitisation will benefit all and improve access to care, education, culture, and all the life improvements the industrial era promised, cannot be envisioned today any more. The gap between cost and benefit, especially of digital services, appears all the more cruelly in events where the victims of floods, rising sea-levels, fires, storms, tempests, and poor harvests, are displayed on cell phones, screens displaying information, and other digital media.[18] Digital media keep us informed but they have also become, in their energetic overshoot, part of what causes the problem, as is shown by studies on the growing energy needs, especially in the domain of the cell phone and internet connection.[19]

---

18   On the connection between socio-economic mechanisms, especially connected to information dissemination, and digitisation, and on solutions to improve current problems, see Lange and Santarius, *Smarte grüne Welt* [74]. The French Agency for Ecological Transition ADEME has developed four scenarios to achieve carbon neutrality in France by 2050. Only one ("pari réparateur") relies heavily on digital technologies as we know them.

19   The development of 5G is a keystone in the report by the Shift Project on the environmental impact of IT and their recommendations for a more sustainable digital future at EU level: https://theshiftproject.org/article/impact-environnemental-du-numerique-5g-nouvelle-etude-du-shift/.

In this challenging context, access to text is an individual issue as well as a societal one and, for some, it is also embedded in professional choices. As a reader, you can choose to buy an analogue book or to purchase an e-reader. A first step to understanding the implications would be for any reader to be able to gain a general sense of the environmental impact of these choices.

But these impacts are all the more difficult to explain unambiguously as they are embedded in national and global structures that add up to more than the sum of individual choices. Economical mechanisms, social relationships, existing (or non-existing) infrastructure, and the weight of political decisions past and present — all of these are at play, intertwined with one another.[20] From the moment when you are part of a society, you have an environmental impact. Social determinism is not the most consoling and satisfying thought here. Would it not be better to cease all activity, meaning, in the case that I am talking about here, to stop reading books altogether to substantially minimise our environmental impact? It is an argument that can be applied similarly to other activities, such as the use of a car, or of a computer — it extends to any human-made artefact.

I would like to draw a parallel with a historical situation in which it was not the environmental cost, but the financial cost that kept readers from buying books.[21] In the late 18th century, literacy goals had born fruit and a much wider array of the population was now able to read. In urban contexts, the cultural capital represented by the knowledge of texts one had read was socially of great interest and potentially a door to better socio-economic conditions. This educative and social validation of literacy led more and more people to gain interest in books and the press. These were not people who wanted to possess books in order to show off, as could be the case in aristocratic milieus, but people who wanted to read translations of the latest novels, practical advice, ideas about hygiene, or poetry. To them, however, printed press and printed books were extremely expensive compared to their income. Several systems were set up to share the cost. In some cases, several people took one subscription to a journal and shared it for reading; then each one of

---

20   See Charbonnier, *Abondance* [36].
21   The following draws from chapter II.5, "Der literarische Markt: Genese, Strukturen, Funktion /Das Publikum" in Kiesel and Münch, *Gesellschaft und Literatur* [72].

them in turn got to keep a copy. Depending on how many people were contributing, it could be every other issue or every third, fourth, fifth, etc., issue. In other cases, they paid a weekly or monthly fee that authorised them to consult and read, in a dedicated room, freshly published items. Some of these library systems were efficient business models for the organising entrepreneur;[22] some were more self-organised by people who put their minimal savings together in order to follow the feuilleton-based adventures of their favourite heroes. But all in all, the trick was to split and share.

While taxing products and services according to their environmental impact could be an interesting experiment to address the issue of reducing the proliferation of greenhouse-gas-emitting artefacts and services — if applied systematically and fairly — such a measure remains to this day out of reach politically and socially on a large scale in the most polluting countries. There is no alternative but to come up with other ways to encourage practices that limit greenhouse gas emissions, perhaps taking inspiration from 18th-century reading circles when it comes to cultural artefacts like text-based media. Without going so far as to nationalise all services, the rule of thumb to minimise impact could simply be that the greater the number of people benefiting from an artefact, the less impactful it is for its single use. Borrowing a book from a friend or a library, or sharing a downloaded digital resource locally are all gestures of reuse that minimise the individual environmental cost for using the concerned item. The production, use, and end-of-life impact can be split among all those who benefit from it, and the part each individual has to account for is reduced.

Sharing is key, and there are ways to make sharing better than it is.[23] I will not explore the economic leverage mechanisms readers can deploy at an individual level to pressure the book market into improving sharing mechanisms, as interesting as this approach may be. Thinking about leverage on environmentally friendly access to text, one central entry point for the development of good practices — and one that perhaps deserves more consideration — is on the part of those who produce text, rather than those who consume it. Let us turn now to those actors for

---

22   A good example is presented in Busch, *Lesezimmer* [32].
23   This aspect is also key in the recommendations made by Lange and Santarius in *Smarte grüne Welt* [74].

whom access to text is not simply a cultural leisure activity, but the core of their professional practice. What take do archivists, publishers, and editors have on the environmental impact of their professional activity?[24]

Wanting to reduce one's environmental impact means striving for a greater energy sobriety. There are several ways to improve things that can be combined differently, depending on the goal. If the goal is to reduce impactful emissions to zero (which would be the basis of what is called carbon neutrality), activity has to cease altogether. Offsetting by planting trees will never fully account for the impact of a digital service: the trade-off of offsetting might make a plausible argument in some areas, based on greenhouse-gas-emissions calculations, but it cannot compensate for larger losses like those in biodiversity or in water resources. From that point of view, the only way to be sure to emit as little as possible is to do nothing. As tempting as this radical option may seem, I will look into less efficient, but less disruptive alternatives.

I will presume that mankind does not purposefully want to self-annihilate in the near future, but strives to pursue something like meaningful human existence at a global level, seeking a form of collective life where not everything is about survival, but where culture at large is part of social cement. For archivists, librarians, publishers, and editors to conceive their activity in such a way that it does as little natural harm as possible, for the largest possible cultural good, it means paying close attention to at least three elements: natural resources, human activity, and energy consumption.

In order to improve the assessment of natural resources consumption in this context, we would need a much greater transparency in information. It is striking how a lack of precise information is the common denominator in the literature dedicated to assessment analyses. This means that in a first step, energy (meaning money, personnel, and actual energy) should be invested in making information on machine components and information transmission infrastructure easily available. In the case of public actors such as archives, libraries, or research institutions, public markets for purchases should take these elements into consideration, offer long warranties, encourage reparability, and provide spare

---

24   In Germany, the *Netzwerk Grüne Bibliothek* has been active for a few years already and provides expertise on sustainable models. They also supply a related bibliography: `https://www.netzwerk-gruene-bibliothek.de/bibliografie/`.

parts and repair. Training public servants to repair the devices they are using in everyday life, or at least to identify what is to be repaired and providing workshops to do so, would, on the one hand, reduce material consumption and spare natural resources and, on the other hand, modify the way we think about our material environment. This could be one of the most important shifts I can think of: not to consider that we are entitled to surround ourselves with tools and services, but to acknowledge the support they provide. This means moving from being annoyed by the cable that does not work anymore to taking good care of one's cables in the first place, looking into repairing and recycling options, and, if it is absolutely necessary to purchase a new cable, then buying one that was produced in conditions that are environmentally acceptable, or checking the availability of a second-hand cable instead. All in all, this means dedicating much more time to the materiality of our environment than we are used to in north-western countries.

More broadly, the notion we have of the time we dedicate to a productive activity with an environmentally impactful output needs a reassessment. As the example above suggests, care for the materiality that surrounds us is bound to take quite some time, especially in the immediate future, considering we have no simple way of gathering the information needed on the impact of our devices, and no simple way to compare options in terms of their environmental and ethical impact. Pioneer work is still required. Reducing the time dedicated to an impactful productive output has an added advantage of making more time available for activities like gardening, barter, craft, and other socio-cultural activities that can be recentred at a more local level, and contribute to lowering overall impact. They would lower overall energy consumption too.

How, in such a setting where the tendency would be to reduce activity, would we manage to guarantee long-term access to textual material? In this field too, energy-saving measures have to be taken. The best measure would be to start before it becomes too complicated — before it is too late and archivists and librarians actually have to choose between cooling the stock room, cooling the server room, or cooling the consultation room when energy shortages and heat waves have become our daily reality and it is not possible to cool all three rooms to counter the 45-degrees Celsius outside temperature. In some respects, optimising existing technologies can make a big difference. In others, we will have to make choices. It will not be possible to archive everything (not that it was possible before, but digitisation might have given the illusion that it did), and it will not be possible to archive in as inflationary a manner as we have done over the past decades. Choices will have to be made, and the criteria for these choices will have to be defined. As I have shown in section 1.1.1, choices in preservation strategies are a highly political issue. However, some of the parameters that come into consideration are not political, but technical.

The many representations of digital text that exist, such as image, raw text, annotated text, metadata, etc., come in different formats. Each different format has a different environmental impact. Simple text like that in metadata or in raw text files is materially close to insignificant compared to a high-resolution scan. Visualisations based on complex calculations are also much more energy-intensive than raw text. Reduced to its duration of life in correct preservation conditions, paper is not necessarily the worst option. At this point, the most efficient way to keep a trace of a textual document may be to provide rich metadata (an index in a digital format that is compatible with other formats) and raw text. The metadata can include information on the materiality of the text and contribute to not losing completely the formal dimension for the sake of preserving semantics. But to be perfectly honest, while we have reliable experience when it comes to preserving paper for several hundred years, our projections are much more speculative when it comes to digital-based formats. A minimal set of purely textual digital information for born-digital documents, in combination with a paper support as far as it exists, could make the core of a sober approach to archiving and making available textual heritage in at least some quantity and some quality. What desirable quantity and quality are remains to be defined more precisely. But I would argue that a sensible approach based on these principles would make it possible to pursue both popular heritage transmission and scholarly work.

Anything that goes beyond this core information should be done in a computer language that is compatible with other languages and that would make it possible to version the file in the manner that was described in section 1.1.2: archiving the basic text in an economical manner, and simply recording changes made over time after that. By proceeding in this way, the virtual space required for archiving remains limited, especially compared to current practices where archives often procure a high-quality scan in TIFF, a high-quality scan in JPG, a lesser quality scan, and a thumbnail of the same manuscript page. Sometimes they even reload all images on each archiving iteration, including those that have not changed in between. There is certainly room for improvement in the current processes.

Procuring the files is one thing, preserving them requires infrastructures. The more institutions and actors share infrastructures, the less environmentally costly they are. Much of the impact of preservation strategies will come down to the way in which datacentres are built, and to their file exchange protocols. For their virtual stock to be shareable, large infrastructures could rely on networks of information that can harvest information as well as distribute it. These networks may be virtual for the users, but they are based on actual cable infrastructure that needs, again, to be conceived in such a way as to optimise the circulation of data. This concerns not only the cables and their dispatch form, but also the type of data they have to transmit. Improving environmental impact can mean favouring some formats over others, such as those that need less bandwidth. On the receiver end, too, energy saving can mean that only the lightest data is transmitted. What happens when you have low bandwidth and webpages load very slowly? This experience gives a sense of all the energy-intensive and superfluous information that is being transmitted with each and every internet request: banners, colours, animations, and videos that start automatically as soon as the page opens.

What is true for infrastructures and for internet protocol is also true for what I would call editorial information. If we want to be able to preserve more than the raw text and a description of page, paper, ink, and writer, this needs to happen in a standardised manner, so that as little energy as possible is used to convert the encoded information and make it legible in various computer systems. We could, for what I have called inherent history and geography of the text in section 1.2.1 for instance, define a set of the information that is likely to be relevant for almost any text, or at least literary text, and agree on a stable way to represent this information digitally. It is fortunate that this is what the TEI consortium has been doing for the past decades already, procuring a solid basis for even complex textual phenomena. Going one step further, the implementation of this standardised, stable, economic way to provide textual and meta-textual information can also be standardised in terms of the workflows it is integrated into. Here, too, editors worldwide could work out a way to arrange the essential digital building blocks that make it possible to progressively enrich text and preserve it in a progressively enriched and enrichable form, thus guaranteeing the availability of basic information and making additions possible. This would mean that archives, libraries, publishers, and scholars all somehow work with a similar, standardised, economical workflow.[25] Of course, this bears the risk of losing information because of its standardisation. But considering today's competitive situation where binary formats jockey for attention, I do not think it would be worse than the loss we would be faced with when we need to save electricity, and datacentres have to be turned off.

Even in a situation where we would have reduced usage to a minimum and saved as much energy as possible in procuring the data basis for textual heritage, the question of duplication remains a crux. Duplication means that datasets are being archived in at least two different locations that mirror one another. If one of the locations ceases to work, or burns down, or if its hard drive content gets erased or crashes, the other iteration can provide backup. Relying on one single copy of digital files is a risky business. But the environmental cost of multiplying by two — if not even by three for a backup of the backup, as is often done — comes down to asking the canon question anew. What exactly justifies a text being preserved in not two, but three high quality copies, in order to be sure, absolutely sure, that it will not be erased from memory? How much are these scans of manuscripts worth for mankind that archivists, librarians, and scholars try to guarantee they will never be subjected to the fundamental rule of any archive, which is that loss and destruction are unavoidable, are part of the process, and have to be accepted as the epistemological premise of all archiving?

In this case, it seems that it is the quantitative dimension (the weighty scans, duplicated two or threefold) that serves as an affirmation of cultural superiority. Yet impact could be more strongly determined at a quality level. Taking technical decisions on formats is also a way of aiding

---

25   In *TEI Models* [16], Julie Giovacchini and I propose a TEI-based approach to reviewing
      and copyediting processes. Ultimately, the goal would be to strive for an even wider
      generalisation.

selection processes. In this type of selection processes too, politics play a part that is not so different from that of 19th-century Germany erecting a cult house to Goethe and Schiller. Topics are prioritised by ministries, funding is made available for these topics, and the amount of funding determines technical choices, and with them the sustainability of the textual resources that will be procured. In this case, much is in the hands of establishment. At best, a small group of educated experts can formulate recommendations. But it has hardly gained transparency in the selection criteria, and is still dominated, at a global level, by English-speaking, educated white male production.

I see a convergence between this form of (political) control on textual content and control on the means of dissemination. While shared infrastructures offer the best guarantee for sustainable preservation, they should ideally rely more on distributed community-based needs and solutions, and not unilateral benefit resulting from top-down instructions. Infrastructures providing long-term hosting of textual heritage should be able to serve as the backbone for initiatives that have little to no means: for instance, a low-cost Raspberry Pi computer and some manpower, with very parsimonious resources, such as those powered by solar panels that are only accessible when the sun is shining.[26] In fact, making digital resources available in different forms depending on an energy scale defined by current physical conditions could be an interesting direction to think about. Instead of making everything accessible all the time, core information could be accessible at all times, and additional information only when renewable energy is available. This would generate new hierarchies between what is deemed indispensable and what is secondary, but at least there would be a coherence, a logic behind availability schemes.

Who, though, would understand that logic? How can it be made comprehensible to users and readers? This question is not only key when it comes to retrieving digitally archived textual material in the long run, but more generally for all the challenges a shift in text access practices is bringing. Current practices have spoiled users to the point that any reader can have the illusion that a plain text search in a browser will open the door to whatever it is they are looking for. This is far from true, as I explained in section 2.2.1, but this misconception is not likely to

---

26 See `https://solar.lowtechmagazine.com/`.

disappear soon. Instead of fostering the delusion that it suffices to know a title and/or an author to find a text, it would make sense to develop educational schemes that provide training in the skills necessary to gain orientation for a digital environment in a context of shortness of resources and ecological mindfulness.

The education I think of would entail a basic training in environmental awareness (covering, among others, the topics I presented in section 3.1.1). It would also train in code and programming literacy, empowering students to be able not only to read and unpack, but also to assess the relevance of computer language choices in the different settings they might encounter. Guidance in heritage selection mechanisms would also be part of it. I wish that my students and children will know better than me how to read XML, how to use a Raspberry Pi, how to work with minimal computing features, and how to manage a simple database. At the time of writing, this type of training is reserved to a handful of ICT students. To me, this type of training is what philology for tomorrow should entail. These are the skills philologists need to develop if we want to have a chance to build our school and higher education curricula on more than a handful of random Google Books.[27]

I strongly believe in educational schemes carried out by professional institutions like state schools or publicly funded universities. But this is perhaps too restrictive, and so is thinking that it is up to the next generation to carry out the change. Maybe this kind of training should be developed on a more widely distributed and accessible level, such as community colleges or *universités populaires*, for people of all ages as long as they understand what is at stake. Maybe it is the wisdom of the masses that will help us renew the canon and keep textual heritage alive, even if only on the days that solar panels can provide energy.

3.2 Trying to make this book an environmental lightweight

The comprehensive character of environmental issues, and the extent to which they are embedded in social processes at large, means that

---

27 I am aware that this is more a concept than an actual training scheme. It leaves unquestioned, as the rest of my argumentation does, the fact that spending a lot of time with one's eyes locked on screens is problematic. What digitality does to bodies is certainly another aspect that needs to be taken into account in this discussion.

addressing them requires, as I suggested above, also a comprehensive response. Material production, individual activity, infrastructure, and workflows need at least to be adapted, if not revised in depth. In the face of all the necessary changes, it remains difficult for environmentally aware individuals to reconsider their own activity in such a way that they do not feel as though they are encouraging inadequate advances, or even develop a sense of guilt about having any activity. The anxiety generated by the sole effort of scrutinising and measuring each of one's step in the world can easily become paralysing.[28]

While it is not the purpose of this book to encourage readers to measure every single one of their activities in the light of its environmental impact, I would like to try to give one concrete example in this last section. My goal is to list the elements that should be taken into account in order to assess the environmental footprint of the conception, production, distribution, archiving, and use of the book you are currently reading. This approach was greatly facilitated by the publishing house, Open Book Publishers, who contributed essential information to the following pages.[29] By diving more into this self-reflexive case study, I intend not only to give a sense of the type of analysis that is necessary to tackle the challenge of initiating practical shifts in key areas. I also want to outline what a coordinated approach, involving the wealth of actors that have leveraging potential to shape access to text for tomorrow, could look like, on the basis of contributions from our current period, which is one of transition.

I begin by presenting the production phase of the book, starting with my own work processes and including those of the publisher. I then move to distribution and archiving strategies, and strive to consider reader behaviour as well.

3.2.1 Writing, printing

At the risk of disappointing readers, I must confess that I hardly used any paper and pen to write this book. This does not necessarily mean

---

28 See Panu Pihkala's synthesis on *Eco-anxiety* [88].

29 I would like to thank Open Book Publishers for their support in this endeavour that required an unusual transparency about internal work processes. In particular, I am grateful to Rupert Gatti for communicating internal documents and information, and engaging in an extensive discussion with me on different aspects of the argument developed in this last section.

that the definition of the onset of the writing process is any easier to identify. In fact, chapters 1 and 2 are largely inspired by earlier publications I wrote and disseminated in a variety of ways over the past ten to twenty years. Most of them have in common that they are preserved and accessible on the online archive HAL,[30] except for the digital scholarly edition *Letters and texts. Intellectual Berlin around 1800*, which was first hosted by the German Trier Center for Digital Humanities,[31] and whose long-term archiving is now ensured by French Research Infrastructure Huma-Num.[32] The preservation and long-term dissemination of these earlier drafts rely on shared public infrastructures.

Both my own earlier publications and the bibliographical information I refer to, chiefly in my footnotes, build a network of explicit intertextuality. In the course of the writing process, I had to check page numbers for quotations or relevant passages, I had to ascertain the wording of citations, and to confirm publication dates. This type of bibliographical quality insurance is a requirement for scholarly work. It involves time and effort in addition to the writing process itself. I did not have to go the library very often since most of the references I had been using were also part of earlier publications I had already been consulting. Library visits for the purpose of writing this book only relate to my Oxford stay in the spring of 2022. From my home in France, I took the train to Oxford and, from my Oxford home, I walked to the different libraries I visited (Taylorian and Library of the Maison Française). It remained rather low-key in terms of impact compared, for instance, to my early career stays in a variety of archives and libraries around the world.

Apart from these visits to actual libraries, I also had to consult references in online libraries and archives for additional details. I usually used meta-catalogues that I already knew well, so that I did not lose too much time browsing the web. When a resource was behind a paywall and inaccessible via my university portal or other libraries I am a member of, I had to resort to other freely accessible resources used by scholars. Assessing the impact of this online activity requires me diving into the footprint of libraries (virtual and physical) and download platforms, their preservation and distribution policies, and each reader's strategy

---

30 See `https://hal.archives-ouvertes.fr/`.
31 See `https://tcdh.uni-trier.de/en`.
32 See `https://www.huma-num.fr/`.

once they are in possession of a copy. We will see later in section 3.2.2 what that actually entails, taking only this book into account.

Evaluating the preparation of chapter 3 added another dimension. I have not been trained in environmental questions in the same manner as in philological ones. I studied philology, wrote a related PhD and a habilitation, and I have been preparing editions for twenty years. My digital training was integrated into my research activity over the past ten years. It took the form of actual trainings in classes, workshops, and hands-on sessions, albeit not as systematic as studying from the onset. Yet, for both the philological and the digital dimensions of this book, I can refer to a classical publication and citation setting, and to an established disciplinary frame of reference, as the bibliography shows. When it comes to environmental questions, my training has been much less systematic — this chapter bears obvious marks of this difference in training quality, especially in the references that frame it. This has to do with the fact that there exists no explicit discipline dedicated to the environmental footprint of dissemination and preservation activities that could be actionable within the research area of Humanities disciplines. The French network Labos 1point5 strives to establish such a field for research activities in general, trying to extend beyond disciplinary boundaries.[33] A large part of what I consider my training in this area consists of interacting with scholars from various disciplines (geography, environment studies, physics, astrophysics, computer science) over the past two years. Yet, to this day, the literature that is discussed and produced in the context of this research network does not address publication issues or access to text at large.

Another French research network, Ecoinfo, is specifically dedicated to tackling the impact of digital technologies, encompassing research activities but not limited to them.[34] Ecoinfo understands itself as a provider of expertise: members benefit from each other's knowledge and experience, and can be trained through interaction and lectures provided within the network. I was lucky to be offered tailored training units by specialists. What I learned about lifecycle assessments — not simply the technical details, but the philosophy of their conception and use — comes out of these e-mail exchanges, training sessions, and discussions in videoconferences.

---

33 See https://labos1point5.org/.
34 See https://ecoinfo.cnrs.fr/.

Calculating the environmental footprint of videoconferencing and comparing the different providers is a complex endeavour.[35] There remain so many uncertainties in how to measure and/or model the components involved in this process (from the devices used, to the internet connection, location of servers, image quality, etc.) that it is only possible to provide scales of magnitude rather than precise figures when it comes to calculating the environmental footprint of videoconferencing. This, again, makes comparisons between different software and systems difficult. As long as it is not yet good practice to provide a numerical assessment of any digital service within the service itself, this kind of endeavours will have limited accuracy. On the one hand, this is annoying because it prevents one from providing a clear assessment. On the other hand, the order of magnitude should actually suffice to raise awareness of the overall necessity to reduce digital activity on a massive scale.

Turning back to my initial question concerning the evaluation of the environmental footprint of zoom training sessions relative to in-person workshops, for instance, the comparison requires one to balance uneven elements. In-person workshops involve transportation, buildings, and material, not at all an insignificant footprint. But they also mean improved communication and greater social well-being compared to videoconferences. For videoconferences just as for in-person training, however, one way to balance human activity and preservation of the environmental footprint could be to build distributed networks of competence that make it possible to disseminate knowledge locally through a pool of trained facilitators. This is precisely what networks of competence are doing: while the environmental impact of the training sessions is *per se* rather high, it has the potential to achieve much improvement at a more local level, through trained people. The more people benefit from it at the end of the training chain, the less the overall environmental impact of the initial training session.

The training I received in zoom sessions is also difficult to evaluate in terms of the bundled competence I have benefited from. Networks of higher education and research professionals are composed of highly trained experts. In the case of the two French networks I mentioned, they

---

35   See the Labos 1point5 and Ecoinfo paper comparing the impact of in-person attendance and videoconference participation at conferences: `https://labos1point5.org/les-infographies/poster-ecoinfo-method`.

bring together the only specialists that exist in small fields that are not established disciplines in and of themselves. The footprint I am trying to optimise here leads me to questions regarding the flexibility of the academic system: what is its ability to make way for emerging disciplinary relevance and integrate it into scholarly discourses, in a context where reference to existing knowledge does not only take the form of books or articles but potentially also podcasts, videos, and executables. These dynamic digital resources present even more citation and copyright issues to deal with than is the case with text, involving the challenges that I mentioned in section 2.2.1. The environmental impact of the production of this book takes into account a fragment of each of these training settings, and of each of these online resources, in the sense that they were seminal to the content I present here, in the book form in which I have brought them together.

Looking now not at the background for the content, but at the technical equipment I have used, the situation is somewhat more straightforward. I wrote this book on a MacBook Air that was purchased by my university in 2018, which I had been using for four years by the time of writing on my computer. I started writing in April 2022 and finished in early September 2022. This initial writing phase was followed by copyediting phases in February-March 2023 and May 2023. On the days I dedicated to it (seldom more than two full days in a week), I spent about eight hours a day writing. I was online most of the time I wrote, and listening to music about half of the time. Using LaTeX, I compiled the document (a single LaTeX file) several times in each hour I spent working on the manuscript. On the days I was working on it, I saved a copy of each daily iteration on an external hard drive, as well as on my University cloud, making one transfer per day on each of the writing days. I have printed the manuscript five times, each time to integrate major edits by third parties or myself, and discussed it over the phone or on zoom meetings for a limited time (two to three hours). If I add up all of this to calculate the overall footprint of the book manuscript, I have to factor the lifecycle assessment of my laptop according to the settings I have in place (screen brightness, screensaver, etc.), of my headset for listening to music, of my hard drive and the University cloud, of internet connections, printing devices from the copy shop, including ink and paper, the consequences of the choice of such a text editor as LaTeX or such a music provider as

Deezer, and the overall electricity consumption during all these activities. This can be done, but needs to combine information that can be measured rather precisely, such as electricity consumption, with elements for which finding the exact information for the exact device I have been using it is unlikely — for instance for my headset, which is a basic model I bought at a railway station a couple of years ago.

I did not make efforts to systematically avoid being online while I was writing, for instance. My goal was not to limit my writing endeavour in order to make it dramatically less impactful; it was rather to see how I could keep the free writing process of my earlier periods of academic activity while trying to limit environmental impact. I did avoid keeping multiple tabs open and limited my dictionary use to three browser tabs (German-English, French-English, Merriam Webster), and I listened to a downloaded playlist whenever I could. These small gestures are really minimal in the big picture, and I consider them minor constraints, just as every historical period has their constraints framing the use of novel technologies.

The more complex the media, the more impact, and the more complex and impactful the archiving process. I applied this principle to the choice of media I used in the book. While I could have integrated graphs, illustrations, images, and colour, these would all have required resorting to more technologies and material at publication, dissemination, and archiving level. This is the reason why this book does not contain anything other than linear text. I purposefully chose not to integrate colours, pictures, or graphs. This required more effort than downloading music playlists. In sections 1.1.2 and 2.2, it would have been much easier to explain the complexity of the different layers of representation involved in digital approaches to text with the help of a few illustrations. I considered the effort of giving up on illustrations worth while not only in terms of environmental impact, but also in terms of inclusivity. Text can always be transposed in audio for the visually impaired, while an image cannot.

The final manuscript for publication presents itself in the form of a PDF generated from the source LaTeX file. During the last month of the writing phase, I interacted with the publisher to discuss copyright and funding, but also layout aspects (so that the final PDF would follow requirements), as well as the content of this section of the book. I also consulted online resources the publisher pointed to. In September 2022, the manuscript went to the publisher who initiated the peer-review process. The book proposal was sent out to three readers who agreed to undertake the review. Some e-mail communication with PDF attachments was involved there. The peer reviewers then sent their reviews back. The commission editor at Open Book Publishers synthetised the reviews and sent them to me, via e-mail again.The revised manuscript was then sent back to reviewers alongside with a list of the revisions, adding another iteration to the process. Overall, this work step required reviewers to be equipped with a basic digital infrastructure: an end device, whether laptop or tablet, with a PDF-reading software and an e-mail programme.

Once the peer review process was finished, the manuscript was sent back to me. I integrated reviewer comments, adapted my LaTeX file in order to comply with the layout requirements that facilitate the identification of the publishing house, and sent back my final PDF to the publisher. Once the edited manuscript was accepted, I sent it to professional copyediting, integrated the copyeditor's edits, and then sent the final version of the manuscript to the publisher.[36] It was the publisher's turn to take a final look at the book as an editorial product. In the case of a PDF generated from a LaTeX file, there is no simple change tracking mode for the publisher to edit the manuscript. We exchanged PDF files.

The final PDF was generated by adding imprints and creating two covers, one for paperback and one for hardback editions. The cover files are created at the publisher's end when the number of pages is final, with the help of a design software. The cover is then integrated into the LaTeX file together with imprint information procured by the publisher. It serves as a basis for generating the e-pub version, which can then work as a pivot for the transformation into alternative formats such as those required by commercial distributors.[37]

The printed book is produced by Ingram, an American company that provides books on demand, based on a service called Lightning Source. Obviously, shipping the volumes from another continent is not really optimal for me as a European author, but considering distribution is not to be limited to my own country, the notion of the centrality of the printing process is a relative one. Emerging models of cradle-to-cradle presses

---

36   I am immensely grateful to Elizabeth Rankin for her magnificent improvements of my text during the copyedit phase.

37   Section 3.2.2 elaborates on this aspect.

focus on using certified material whose impact is maximally lowered, but this model is not implemented widely enough to be recognised as a solid one for scientific publishers with specific market requirements.[38]

In the phase of conception, preparation, and production of the output PDF file, there is not much room for alternative processes; there is at any rate no easy and obvious way to optimise them. Using an open-source software like LaTeX rather than a commercial text editor already modifies substantially the workflows compared to what publishers are used to. There exists no standardised way of writing, reviewing, editing, and preparing for a generic representation even of scientific texts today. Several types of software, formats, and processes come into play, which may be more or less open, and more or less compatible. Moving to a radically more environmentally friendly process would require to introduce profound changes of habits in a field where it is already difficult for small publishers to find an economically viable balance. In terms of the pressure imposed by the market on the publisher, I have remained, as an author, free to make a range of decisions and, since I am curating the source file, I retain complete control over the text document.

Much of the impact related to the production of this book has to do not so much with the production itself, but with the next step: its dissemination to an audience. In the next section, I look into distribution circuits for the published book and its archiving for later consultation.

3.2.2 Distributing, archiving — and the readers

Being able to actually hold the printed book in my hands is certainly satisfactory. Knowing it has been made possible without contributing too massively to processes that undermine the preservation of natural resources even more so. Only a few copies have to be printed for traditional library distribution, to be sent to the UK branch of Gobi (Global Online Bibliographic Information), and a few more for selected journals that are likely to commission reviews. All other prints will be on demand, when ordered by either libraries, bookstores or readers.

But if the goal was to reach an audience as wide as possible with as limited an impact as possible — and it is — the book artefact is only a

---

38 In *Zwischen Resilienz*, Wittenbrink presents an experiment with cradle-to-cradle presses and explains its mechanisms; see [103].

nice by-product. Free online access can reach many more people. Yet the extent of its success at reaching a wide array of people depends on the form of publishing and the type of dissemination involved.

I could have chosen to self-publish this book. As I have explained in section 2.2.2, self-publishing has become a fairly straightforward endeavour in the digital context. To publish this book, I could have set up a webpage, written a series of blogposts, or simply have deposited it in a pre-print archive. I could have made the text available for peer review using an open peer review platform if I wanted to integrate some form of quality control.[39] Citability could have been guaranteed through the stable URLs provided by the infrastructures hosting open pre-print repositories or scholarly blogs, depending on the form I had chosen, which also guarantee long-term archiving in the output format.[40] Since the only feature is structured text, layout would not have been much of a problem.

But I would have had to maintain the webpage or blog, or rely on the platform I used to do it, in order to guarantee access to my text. When relying on an online framework provider for publication, one does not really have a say in the preservation and access strategies pursued by the provider. While the publication process would be taken care of via an existing technological solution, and while generic harvesting would improve findability of the text compared to, for instance, a basic print version, it would be fully up to me, the author, to arrange for the text to make its way to its audience. Such input as professional proofreading and editing, layout instructions, and audience access would be limited. Nor would it provide a clear notion of the environmental impact of the deployed technological solution. Working with a publisher facilitates all of this, and working with Open Book Publishers has the advantage of bringing an exceptional transparency and a reliable technical quality assessment to the process. Also, since I am not transferring the rights on my work, which is published under a CC-BY license (requiring the author to be named in case of citation or reuse, but without additional restrictions to reuse), there is nothing to prevent me from engaging into

---

39   For instance `https://web.hypothes.is/`.

40   The scholarly blogging platform `https://hypotheses.org/` provides stable URLs and ISSN numbers in coordination with the French Bibliothèque Nationale; the preprint archive HAL `https://doc.archives-ouvertes.fr/en/homepage/` also provides stable URLs for each version of a scholarly work they store.

self-managed dissemination in addition to the professional distribution provided by the publisher.

Distribution of the digital book by Open Book Publishers takes place primarily via their website,[41] where the PDF and the e-pub of the book are hosted under a dedicated URL and referenced through a unique DOI. They are accessible without charge. The webpage of Open Book Publishers is kept minimal in terms of design, so that download time, and hence the energy required for download, is curtailed.[42] From there, distribution is extended to what could be considered mega-catalogues, which consist of metadata aggregators that are either library-based or connected to larger infrastructure projects. These providers use the metadata, but they rely fully on the Open Book Publishers website and link to it. These massive databases aggregate metadata of Open Access resources, making them more easily findable than they would be if they were listed only on the publisher's website.[43] This remains fairly economical as long as it is not excessively multiplied (internal documentation points to about ten such metadata aggregators for Open Book Publishers) and as long as metadata can be exchanged in a standardised manner. In any case, metadata transit remains a not too costly process environmentally compared to content exchange.

The situation is more worrisome when it comes to the dissemination via digital book distributors, who gather and distribute not only metadata, but also content. These commercial actors usually develop their own formats, and the e-pub output has to be transformed again in order to be accessible via Amazon or Google Books. Additionally, they generate their own URL, sometimes even their own DOI, for the same book that is already referenced on the publisher's website. This redundancy is neither good for the environment nor for the advancement of knowledge. References are much more likely to get lost if they are equivocal. As if multiplying home-made formats and DOIs for the same text entity was not bad enough, some of these book distributors apply referencing and formatting at a different level. Some of the DOIs are attributed not to the book as a whole on the publisher's website, but to chapters. This book,

---

41   See https://www.openbookpublishers.com/.
42   It also fulfils accessibility standards.
43   One example of such an aggregator is the Directory of Open Access Books, or DOAB; see https://www.doabooks.org/.

with its six parts (including introduction, conclusion, and bibliography), would and certainly will, at some point, be attributed six different DOIs by one book distributor or the other. This means that there will be one DOI for the book as a whole and six DOIs for each chapter, maybe even several DOIs for either the book or the chapter depending on the distributor's practice. This redundancy issue is not simply problematic in terms of environmental cost or in terms of efficient referencing, it also makes any use of aggregated statistics futile. If some providers work at chapter level and others at book level, it is impossible to get a clear view on what the overall download activity for a single book is. It would be like adding apples and oranges. Yet downloads are a relevant indicator when it comes to assessing environmental impact as it informs on data traffic.

All the previous considerations concern the distribution of the book via online platforms. These distribution networks provide accessibility for a wide array of people with a comparatively limited, although not totally negligible environmental impact. But as I have shown in Chapter 1, online platforms do not necessarily guarantee a book's accessibility in the future. When considering the long-term accessibility of this book, even the combination of library distribution and Open Access availability does not provide a sustainable preservation, let alone access. Supplying archiving requires one to have a way to store and record the text file (as described in section 1.1.2) on a reliable infrastructure. By the time of writing, Open Book Publishers has a two-step archiving process that is based on the PDF output. It is archived by Portico,[44] a service that supplies only archiving. Access is closed, only to be opened if the primary distributor is unable to provide access anymore — if they cease their publishing activity for instance. A PDF is also provided to the Internet Archive, together with all the links included into it in their form at the time of publication. The Wayback Machine thus provides access to the book itself and to the interlinked material. This twofold solution involving Portico and the Internet Archive is based on external services and is not fully reliable in the sense that these services could very well, at one point or another, be deactivated. Then, the better form of archiving will definitely be the traditional library.

No simple way to archive scholarly books exists that would guarantee long time preservation and access, and all of this at a minimal

---

44   See https://www.portico.org/.

environmental cost. Open Books Publishers has been working towards a more sustainable approach in the context of the COPIM project, together with institutional and commercial partners.[45] Instead of working with a PDF-based output format, COPIM is building an infrastructure that will provide long time archiving of XML-TEI files containing raw text, as well as the document structure, such as sections and subsections, reflecting the overall tree structure of the text. The distributed infrastructure underlying the archiving process will be maintained by a network of actors, including long-established university libraries that can guarantee sustainability. Since I am working in TEI in my everyday editorial work and am familiar with this technology, I provided the TEI file generated from the LaTeX file myself, but the publisher could have taken care of it. This conversion does add to the overall footprint, but it does so from a rather parsimonious language to another parsimonious one, and in the case of this book that has very few features and only text, the process is an economical one. The environmental cost is extremely low when measured against the potential archival benefit.

The archiving repository envisioned by the COPIM project provides sustainability at a low ecological cost by building on existing infrastructures. The libraries involved have a long-standing history,[46] and are not likely to disappear from one day to the next. Relying on a durable infrastructure is only one way of keeping the environmental footprint of this archiving strategy fairly low. Information exchange within the network is also kept at a minimum through the use of completely open and interlinkable metadata catalogues. The praise of metadata and catalogues I have been singing all along in this book will not stop at the end of Chapter 3. If anything, it will become even louder. Standardised metadata and catalogues are not just part of an efficient archiving and distribution process, they are also key to a low environmental impact of access to text.

What is more, since the network is community-driven and not commercial, it saves on such energy dispensers as ad banners on websites and, more generally, environmentally costly features designed to increase business wins. With this minimal energy outlay, it also contributes to preserving data that would otherwise be at risk of disappearing. It is

---

45   See Community-led Open Publication Infrastructures for Monographs, `https://www.copim.ac.uk/`.

46   See the list here: `https://www.copim.ac.uk/about-us/who/`.

tailored for and by small publishers, who are more likely than not to cease their business long before the partner infrastructures and heritage institutions do. In that sense, it contributes to preserve digital material that would otherwise die with its initiating publisher. The COPIM project is still nascent to this day, and has not yet unfolded its full potential. Yet the rationale behind its creation shows, at the very least, that key actors in the field of access to text are not only aware of the issues at stake but have taken concrete action to tackle them. I can only hope that more of these convergences will emerge and foster a constructive dialogue between publishers aiming at scholarly and digital quality and higher education and cultural heritage institutions in the years to come.

With these different work steps, the distribution and archiving of this book are guaranteed along the lines of what is state of the art in the European context at the time I am writing. While it would certainly be possible to further reduce the environmental impact of the production of the book, maybe also to do so in a noticeable manner, this would involve a major disruption in the writing, editing, and publishing processes. The solution implemented for this book displays a good balance between environmental impact, respect of pre-existing working habits, quality insurance mechanisms, solidity of archiving strategies and speed of the overall publication and distribution process. It remains within the realm of what competitive publishing processes impose upon individuals in the academic system — authors, reviewers, editors and publishers — while modifying otherwise environmentally costly items. It has one foot in the old world of speedy digitisation and availability and the other in what one can only hope will be a new world of reduced pace and production with inclusive access practices.

There remain two dimensions of the environmental footprint of this book to discuss. One concerns advertisement strategies, and the other the way in which user behaviours are to be accounted for.

Distribution strategies include the physical dissemination of book copies (packaging, sending, and delivering) and that of the PDF mentioned earlier, with its dispatch on a variety of relevant portals. But this is only a fragment of the overall dissemination activity of a traditional publisher. Traditionally, publishers would be connected to a network of libraries to which they would send their catalogue of new publications, and they would also present these catalogues at major book fairs.

Open Book Publishers does not engage in this kind of practice: it neither publishes a printed catalogue nor sends its representatives to book fairs throughout the world, having a strict no flight policy. Most publishers have not engaged in such low-energy practices; in these cases, additional layers have to be added to the calculation.

Advertisement strategies developed as early as the publishing business itself, but gained new traction from the early years of the 20th century. From the first textual advertisement appearing in chapbooks to radio and television teasers, the placement of cultural products has evolved together with the means of diffusion and communication. Not only commercial advertisement, but also critical discussion and reviewing has moved from private correspondences and arcane columns to television programs discussing new releases, inviting authors, and professionalising critique in cultural production.

In the digital context of the 21st century, it has become standard to present new books on the publisher's webpage, in newsletters, or on mailing lists. In the last decade, social media has also become a major outlet. To some extent, one could consider the web of intertextual references that is generated in this manner as an overarching intertextuality, but it is important to be aware of the commercial dimension this entails. In the late 19th-century novel *Effi Briest*, the eponymous main protagonist hears that her husband has uncovered her unfaithfulness and her parents rejected her in a letter that the author, Theodor Fontane, purposefully places on a table next to a certain shampoo advertisement. This is interpreted as innovative and ultimately, from today's point of view, positive.[47] But when Netflix productions integrate commercial brands on purpose as a form of hidden advertisement, inciting thousands of teenagers to use a product, listen to a band, eat a candy — or read a book — the embedding of commercial and cultural purposes gains a novel dimension, among others in terms of its environmental footprint. This kind of branding adds to the overall impact, as do all dissemination methods used directly or indirectly. There, too, the sum of the impact is not easy to measure since the advertisement for one product is embedded in a complex cultural context, a hidden advertisement in popular by-products.

In terms of the environmental impact of advertisement campaigns, the most environmentally impactful process is obviously one that would in-

---

47   See Lyon, *Anzeigen* [77], pp. 385-386.

volve an upscaled campaign that would not attract the hoped for number of readers. The ratio between the number of readers and the deployed efforts among which the environmental production cost can be split is then very unfavourable. The ability to tailor this ratio with regard to commercial success is precisely the type of competence that publishers provide. It requires an in-depth knowledge of evolving commercial mechanisms — but also an ethical view on them. In the context of the climate crisis, publishers will need to add to this competence a sense of the ecological impact of the processes involved, and find viable ways to navigate between contradictory requirements. In this regard, Open Book Publishers remains rather restrained, having no dedicated commercial strategy for distribution. It relies on twitter posts and on the author's network for one part, and on the discoverability that is made possible through the dissemination of rich metadata via meta-catalogues and platforms for the other. There is no major additional environmental asset to be achieved at that level concerning this book.

In addition to production and distribution activity, the assessment of the environmental footprint of this book requires one to also take into account the receiver's end: the activity generated by the readers. Since the book exists as a printed version and an online version, this means considering the readers of both versions of the book. The more infrastructure is shared by readers, the less the impact of the use. Compared to those who purchase a book for their sole use, readers who borrow the book from a library and then bring it back for others to read, for instance, lower the overall impact: the ratio to be calculated corresponds to the part this book represents in the overall stock supported by the concerned library, divided by the number of users. Calculating the impact of access to the digital version of the book is more complicated since it depends on the end devices and the internet connections used by the readers to download, store, and access the book. If they use reconditioned material and a wired connection, the impact will be lower than if they download the book with a 5G connection on a brand-new smartphone. One could imagine that author and publisher could provide recommendations regarding user behaviour, pointing to optimal settings for accessing the text in a thrifty manner. Superfluous energy spending can be circumvented by keeping the download page as simple as possible, avoiding banners, animations, and data tracking of all sorts, as Open Book Publishers does with its website. But it cannot prevent readers purchasing devices whose environmental production, distribution, use, and end of life cost will skyrocket the overall impact.

This concerns only direct consumers of the book content, and for this book, I can only assume that there will not be no further elements to take into account in the calculation. But in order to be consistent, for a publication that would potentially target a larger audience, for instance, it would also be necessary to consider the impact of the intertextuality generated by the book production, that is, not the users of the book itself, but the consumers of targeted advertisement via other media such as the press, TV, movies, etc. This is a point where it becomes difficult to decide where to draw the line. Do I need to add the impact of the Netflix series in which a book is mentioned if I want to measure the impact of its reception? While such providers are environmentally extremely impactful, it would mean entering the domain of popular culture, and including the production of cultural artefacts that are shared by a very wide array of the population, and contributing, to some extent, to the canon of popular culture. What is the environmental price paid for sharing culture at large? The way cultural and commercial interests are intertwined points to the fact that shifts in the cultural canon induced by digitisation will also have to be reshaped by the consequences the climate crisis exerts on the material conditions of human culture.

I do not have many final conclusions to draw from this attempt to reflect on environmentally aware workflows for providing access to text. As long as we are not able to rely on transparent information, only radical disruptions are likely to activate leverages towards a sustainable future.

Yet, the pessimistic observation that initiated my change of perspective on the role of digital media in accessing text can be reconsidered in the light of this small journey through various assessments. Not all is good in digital technologies, and their being embedded in the socio-economic fabric of highly resourced capitalism does not facilitate unselfish practices. But there is little doubt that sharing and steering towards sustainable infrastructures — solutions that have proved effective in modern contexts — can contribute to alleviating the environmental cost of producing, distributing, enjoying, and preserving access to text.

Conclusion

While I have spent most of my adult life archiving, publishing, writing, and editing, convinced that advances in knowledge are facilitated by digital media, the accelerating climate crisis has moved the parameters of what I have always considered my vocation in a way that is fundamental for the values that are at the heart of my vision of textual studies. The extent to which my inner orientation was shaken by this shift in the way I conceived my mission as a scholar called for a new form of enquiry. For once, I did not publish a blog post, a scholarly article, a series of tweets, or an edition of an old manuscript — in the hope, maybe, of reaching new readers and opening new transmission chains, but also, certainly, to find peace and reconcile with the scholarly choices I had made across the years. There is something contradictory in writing a book about the fact that so much is changing about texts and books that I do not really know how long the book will be there for an audience to read, in what form, or for whom. But it mattered to me to shed light on the consistency between considerations on environmental issues and reflexive work on digital media. In this book, I wanted to show that philology, Open Access, and environmental evolutions are intimately intertwined.

This called for detours. I revisited many of my earlier publications in a condensed form in the first chapters of this book. The European perspective I bring to such topics as archiving, text constitution, or the advent of a publisher-based book market is strongly connected to the objects I know best: early modern manuscripts, scholarly books, and artefacts encapsulating the canons of literary history. Considering them in their digital dimension opened new perspectives on the theories and artefacts of my early career, even more so in the environmental context of more recent years. With my education and my training, I am not able to offer more than what I think remains a traditional European perspective. I am also aware that I have only superficially touched on economic and legal aspects that would add a decisive dimension to the argument, just as my scattered educational suggestions lack a theoretical basis. But for all that is not in this book, I still hope that what there is can open a much

    https://doi.org/10.11647/OBP.0355.04

needed dialogue, and bring to the question of access to text in the context of the climate crisis the attention it deserves.

Considering the opposition between nature and culture has long been a theoretical, if not rhetorical, topic of interest in north-western countries. Technological advances have now become so dominant in the regulation of interpersonal relationships that natural physical time and space tend to lose their mental substance as a framework of reference. Drawing a direct line between elaborate technological artefacts such as digital editions and the natural resources they require in order to be manufactured is not instinctive, nor is it culturally fostered in north-western countries. It was my goal here to uncover all the intermediary steps that digital philology can provide in order to make this process visible and to raise awareness of the emergency there is to show it more clearly. By creating cultural artefacts, we destroy natural resources: this finding is so crushing that it can lead to the conclusion that cultural activities, at large, are destructive — a similar predicament to that of activists spattering soup on Van Gogh's *Sunflowers*. But, like them, I do not want to break the glass protection that preserves the artwork.

With this book, I wanted to show that as a society, even as a global one, we have been able to build the material and intellectual conditions to provide access to community-building cultural artefacts, especially text. It is up to today's actors to turn it into an asset in the context of the climate crisis and to envision a future where access to text is a common good that even more societies can rely on than it is the case today. We know how to make access to text sustainable in the middle and long run with hybrid settings building on physical artefacts and the digital forms of their representation. We know that in order to make it work, we need to offer dedicated training and erect infrastructures to preserve and distribute information that will be accessible to all. And we know there is no time to lose.

Having reached the end of this journey through text, yet another question is still waiting for an answer. What should I do with the old papers I found in the family home? My mother would throw them away, my daughter would want to keep all of them, but she does not know how to do so. It is pretty much up to me to decide, and to choose what remains in the drawer, what goes online, and what will only be a vague memory in the stories I tell my children and hope that they pass them on to the next generation.

Index

Bibliography

[1]   Janneke ADEMA, *Living Books. Experiments in the Posthumanities* (Cambridge, MA: MIT Press, 2021), `https://livingbooks.mitpress.mit.edu/`.

[2]   Sheila ANDERSON, "What are Research Infrastructures?", *International Journal of Humanities and Arts Computing*, 7 (2013), p. 4-23.

[3]   Marguerite AVERY, Alex HOLZMAN and Robert BROWN, eds, *Journal of Scholarly Publishing, Special Issue on Open Access*, 49 (2017), `https://www.utpjournals.press/toc/jsp/49/1`.

[4]   Anne BAILLOT (ed.), *Netzwerke des Wissens. Das intellektuelle Berlin um 1800* (Berlin: Berliner Wissenschaftsverlag, 2011).

[5]   Anne BAILLOT, "Die Entdeckung der Ironie. Das Bild Shakespeares bei Tieck und Solger", in *Shakespeare und kein Ende? Beiträge zur Shakespeare-Rezeption in Deutschland und in Frankreich vom 18. bis 20. Jahrhundert*, ed. by Béatrice Dumiche (Bonn: Romanistischer Verlag, 2012), pp. 81-95.

[6]   Anne BAILLOT, "Wissen, Lieben – und Schreiben: Phantastik und Skepsis im Briefwechsel Chamissos mit seiner Frau aus dem Sommer 1823", in *Phantastik und Skepsis. Adelbert von Chamissos Leben- und Schreibwelten*, ed. by Roland Berbig, Walter Erhart, Monika Sproll and Jutta Weber (Göttingen: Vandenhoeck and Ruprecht, 2016), pp. 351-367.

[7]   Anne BAILLOT, "Was tun mit der Weisheit der Massen? Moderne Philologie im digitalen Zeitalter", in *Symphilologie. Formen der Kooperation in den Geisteswissenschaften*, ed. by Stefanie Stockhorst, Marcel Lepper and Vinzenz Hoppe (Göttingen: Vandenhoeck and Ruprecht, 2016), pp. 261-279.

[8]	Anne Baillot, "Das Netzwerk als Kunstwerk", in *Die Kommunikations-, Wissens-, und Handlungsräume der Henriette Herz (1764–1847)*, ed. by Hannah Lotte Lund, Ulrike Schneider and Ulrike Wels (Göttingen: Vandenhoeck and Ruprecht, 2017), pp. 45-53.

[9]	Anne Baillot, "Die Krux mit dem Netz. Verknüpfung und Visualisierung bei digitalen Briefeditionen", in *Quantitative Ansätze in den Literatur- und Geisteswissenschaften. Systematische und historische Perspektiven*, ed. by Toni Bernhart, Marcus Willand, Sandra Richter and Andrea Albrecht (Berlin: De Gruyter, 2018), pp. 355-370.

[10]	Anne Baillot, "Reconstruire ce qui manque — ou le déconstruire? Approches numériques des sources historiques", *Digital Humanities Quarterly*, 12.1 (2018), `http://digitalhumanities.org:8081/dhq/vol/12/1/000348/000348.html`.

[11]	Anne Baillot, "'Moi, solitaire, tel Merlin depuis son tombeau lumineux, faisant entendre mon écho paisiblement, parfois tout près, occasionnellement au loin.' Le vieux Goethe et la reprise de l'œuvre au miroir de la correspondance", in *Goethe, le second auteur. Actualité d'un inactuel*, ed. by Denis Thouard and Christoph König (Tusson: Hermann, 2022), pp. 207-221.

[12]	Anne Baillot and Anna Busch, "'Berliner Intellektuelle um 1800' als Programm. Über Potential und Grenzen digitalen Edierens", *literaturkritik.de*, Special Issue "Romantik digital" (2014), `http://www.literaturkritik.de/public/rezension.php?rez_id=19678&ausgabe=201409`.

[13]	Anne Baillot and Anna Busch, "Vernetzung—Erzählung—Kollation. Digitale Methoden in der Biographieforschung", *BIOS – Zeitschrift für Biographieforschung, Oral History und Lebensverlaufsanalysen*, 30 (2019), pp. 22-29.

[14]	Anne Baillot and Anna Busch, "Editing for Man and Machine: The Digital Edition *Letters and Texts. Intellectual Berlin around 1800* as an example", *Variants*, 15-16 (2021), `https://doi.org/10.4000/variants.1220.`.

[15]   Anne BAILLOT, Christiane HACKEL and Sabine SEIFERT, "Neue Perspektiven der August Boeckh-Forschung", *Geschichte der Germanistik*, 41/42 (2012), pp. 139-140.

[16]   Anne BAILLOT and Julie GIOVACCHINI, "TEI Models for the Publication of Social Sciences and Humanities Journals: Opportunities, Challenges, and First Steps Toward a Standardized Workflow", *Journal of the Text Encoding Initiative*, 14 (2021), `https://doi.org/10.4000/jtei.3419`.

[17]   Anne BAILLOT, David LASSNER, Julius COBURGER, and Clemens NEUDECKER, "Publishing an OCR Ground Truth Data Set for Reuse in an Unclear Copyright Setting. Two Case Studies with Legal and Technical Solutions to Enable a Collective OCR Ground Truth Data Set Effort", *Fabrikation von Erkenntnis – Experimente in den Digital Humanities*, Special Issue 5 of the *Zeitschrift für digitale Geisteswissenschaften*, ed. by Manuel Burghardt, Lisa Dieckmann, Timo Steyer, Peer Trilcke, Niels Walkowski, Joëlle Weis and Ulrike Wuttke (2021), `https://doi.org/10.17175/sb005_006`.

[18]   Anne BAILLOT and David LASSNER, "Von Graphen zu Word Embeddings — Zur Entwicklung des mathematischen und visuellen Instrumentariums der Literaturwissenschaft", in *Landkarten und Zeitleisten: zur Funktion von Bildern in der Literaturgeschichte*, Special Issue of *Germanica*, ed. by Bénédicte Terrisse and Werner Wögerbauer (2022), `https://doi.org/10.4000/germanica.19002`.

[19]   Anne BAILLOT and Sophia ZEIL, "Tieck und Solger. Zwei Namen und ihre intellektuellen Genealogien", in *Ludwig Tieck. Werk – Familie – Zeitgenossenschaft*, ed. by Achim Hölter and Walter Schmitz (Dresden: Thelem, 2021), pp. 109-142.

[20]   Philip C. BANTIN, "Strategies for Managing Electronic Records: A New Archival Paradigm? An Affirmation of Our Archival Traditions?", *Archival Issues*, 23.1 (1998), pp. 17-34.

[21]   Frédéric BARBIER, *Trois cents ans de librairie et d'imprimerie: Berger-Levrault, 1676-1830* (Geneva: Droz, 1979).

[22]   Roland BARTHES, *Le degré zéro de l'écriture* (Paris: Seuil, 1972).

[23] Syd BAUMAN, "Interchange vs. Interoperability", *Proceedings of Balisage: The Markup Conference 2011*, Balisage Series on Markup Technologies, 7 (2011), https://doi.org/10.4242/BalisageVol7.Bauman01.

[24] Elizabeth BIRD, "Are We All Produsers Now?", *Cultural Studies*, 25 (2011), pp. 502-516.

[25] Roman BLEIER, Martina BÜRGERMEISTER, Helmut W. KLUG, Frederike NEUBER and Gerlinde SCHNEIDER (eds), *Digital Scholarly Editions as Interfaces* (Norderstedt: Schriftenreihe des Instituts für Dokumentologie und Editorik, 2018), https://www.i-d-e.de/publikationen/schriften/bd-12-interfaces/.

[26] August BOECKH, *Encyklopädie und Methodologie der philologischen Wissenschaften*, ed. by Ernst Bratuscheck (Leipzig: Teubner, 1877).

[27] Philipp BÖTTCHER, "Tieck und seine Verleger" in *Ludwig Tieck. Leben — Werk — Wirkung*, ed. by Claudia Stockinger and Stefan Scherer (Berlin: De Gruyter, 2011), pp. 148-164.

[28] Christine L. BORGMAN, *Scholarship in the Digital Age: Information, Infrastructure, and the Internet* (Cambridge, MA: MIT Press, 2007).

[29] Teun BOUSEMA, Leonard BURTSCHER, Ronald P. VAN RIJ, Didier BARRET and Kate WHITFIELD, "The critical role of funders in shrinking the carbon footprint of research", *The Lancet* (2022), https://doi.org/10.1016/S2542-5196(21)00276-X.

[30] Marie BROSIUS (ed.), *Ancient Archives and Archival Traditions: Concepts of Record-keeping in the Ancient World* (Oxford: Oxford University Press, 2003).

[31] Lou BURNARD, *What is the Text Encoding Initiative? How to Add Intelligent Markup to Digital Resources* (Marseilles: OpenEdition Press, 2014), https://doi.org/10.4000/books.oep.426.

[32] Anna BUSCH, "Wissensorganisation und -vermittlung in der Gründungsphase der Berlin Universität. Julius Eduard Hitzigs Lesezimmer für die Universität als erste Berliner Universitätsbibliothek", in *Netzwerke des Wissens*, ed. by Anne Baillot (Berlin: Berliner Wissenschaftsverlag, 2011), pp. 151-164.

[33] Anna Busch, "Visualisierung textgenetischer Phänomene in digitalen Editionen", *Textgenese in der digitalen Edition*, Special Volume of *Editio*, ed. by Anke Bosse and Walter Fanta (Berlin, 2019), pp. 51-63.

[34] Anna Busch and Johannes Görbert, "'Rezensiert und zurechtgeknetet.' Chamissos Briefe von seiner Weltreise – Original und Edition in Gegenüberstellung", in *Phantastik und Skepsis. Adelbert von Chamissos Lebens- und Schreibwelten*, ed. by Roland Berbig, Walter Erhart, Monika Sproll and Jutta Weber (Göttingen: Vandenhoeck and Ruprecht, 2016), pp. 111-144.

[35] Jean-Claude Carrière and Umberto Eco, *N'espérez pas vous débarrasser des livres* (Paris: Grasset, 2009).

[36] Pierre Charbonnier, *Abondance et liberté. Une histoire environnementale des idées politiques* (Paris: La Découverte, 2020).

[37] Roger Chartier and Henri-Jean Martin (eds), *Histoire de l'édition française* (Paris: Promodis, vol. 1, 1982; vol. 2, 1984; vol. 3, 1985; vol. 4, 1986).

[38] Terry Cook, "What is Past is Prologue: A History of Archival Ideas since 1898", *Archivaria*, 43 (Spring 1997), pp. 17-63.

[39] Robert Darnton, "What is the History of Books", *Daedalus*, 111.3 (1982), pp. 65-83.

[40] Robert Darnton, *Censors at Work: How States Shaped Literature* (New York: Norton, 2014).

[41] Jacques Derrida, *De la grammatologie* (Paris: Les Éditions de Minuit, 1967).

[42] Jacques Derrida, *Mal d'archive: une impression freudienne* (Paris: Galilée, 1995).

[43] Jacques Derrida, *Papier Machine. Le ruban de machine à écrire et autres réponses* (Paris: Galilée, 2001).

[44]   Glenn Dingwall, "Digital Preservation: From Possible to Practical", in *Currents of Archival Thinking*, 2nd edition, ed. by Heather Mac Neil and Terry Eastwood (Santa Barbara: Libraries Unlimited, 2017), pp. 135-161.

[45]   Knut Ebeling and Stephan Günzel (eds), *Archivologie. Theorien des Archivs in Philosophie, Medien und Künsten* (Berlin: Kadmos, 2009).

[46]   Jennifer Edmond (ed.), *Digital Technology and The Practices of Humanities Research* (Cambridge: Open Book Publishers, 2020), `https://doi.org/10.11647/OBP.0192`.

[47]   Martin Paul Eve, *Open Access and the Humanities: Contexts, Controversies and the Future* (Cambridge: University Press, 2014), `https://doi.org/10.1017/CBO9781316161012`.

[48]   Arlette Farge, *Le goût de l'archive* (Paris: Seuil, 1989).

[49]   Melanie Feinberg, "Beyond Digital and Physical Objects: The Intellectual Work as a Concept of Interest for HCI", *Proceedings of the SIGCHI Conference on Human Factors in Computing Sysems* (Paris, France, 27 April-2 May, 2013), pp. 3317-3326.

[50]   Bernhard Fischer, *Johann Friedrich Cotta. Verleger — Entrepreneur — Politiker* (Göttingen: Wallstein, 2014).

[51]   Bernhard Fischer and Gabriele Klunkert (eds), *Goethe- und Schiller-Archiv* (Weimar: Klassik Stiftung Weimar, 2012).

[52]   Ernst Fischer, "'Die merkwürdige Verbindung als Freund und Geschäftsmann'. Zur Mikrosoziologie und Mikroökonomie der Autor-Verleger-Beziehung im Spiegel der Briefwechsel", *Leipziger Jahrbuch für Buchgeschichte*, 15 (2006), pp. 245-286.

[53]   Michel Foucault, *L'Archéologie du savoir* (Paris: Gallimard, 1984).

[54]   Matthias Freise, Alexander Starre, Matthias Beilein and Aleida Assmann, "Literaturwissenschaftliche Kanontheorien und Modelle der Kanonbildung", in *Handbuch Kanon und Wertung. Theorien, Instanzen, Geschichte*, ed. by Gabriele Rippl and Simone Winko (Stuttgart: Metzler, 2013), pp. 50-84

[55] Sigmund FREUD, *Das Unbehagen in der Kultur und andere theoretische Schriften* (Frankfurt/Main: Fischer,1994).

[56] Nikolaus GATTER, *"'Lebensbilder, die Zukunft zu bevölkern': von Rahel Levins Salon zur "Sammlung Varnhagen"* (Cologne: Varnhagen Gesellschaft, 2006).

[57] Nikolaus GATTER, *"'Gift, geradezu Gift für das unwissende Publicum': der diaristische Nachlaß von Karl August Varnhagen von Ense und die Polemik gegen Ludmilla Assings Editionen* (Bielefeld: Aisthesis, 1993).

[58] Petra GEHRING, "Archivprobleme", in *Handbuch Archiv*, ed. by Marcel Lepper and Ulrich Raulff (Stuttgart: Metzler, 2016), pp. 17-20.

[59] Jürgen GERHARDS, Astrid EICHHORN, Julia FISCHER, Ute FREVERT and Christoph MARKSCHIES, *Klimaschutz und akademische Dienstreisen. Empfehlungen für ein umweltschonendes Reiseverhalten in der Wissenschaft* (Berlin: Berlin-Brandenburgische Akademie der Wissenschaften, 2022), `https://www.bbaw.de/files-bbaw/publikationen/denkanstoesse/BBAW_Denkanstoesse_9_2022_Lay5_RZ_Web-A.pdf`

[60] Anne J. GILLILAND-SWETLAND, "Electronic Record Management", *Annual Review of Information Science and Technology*, 39.1 (2005), pp. 219-253, `https://doi.org/10.1002/aris.1440390113`.

[61] Anne J. GILLILAND, Sue McKEMMISCH, Andrew J. LAU (eds), *Research in the Archival Multiverse* (Monash: Monash University Publishing, 2017).

[62] Anthony GRAFTON, *La page de l'Antiquité à l'ère du numérique* (Paris: Hazan, 2015).

[63] David C. GREETHAM, *Textual Scholarship* (New York/London: Garland, 1992).

[64] David C. GREETHAM (ed.), *Scholarly Editing: A Guide to Research* (New York: Modern Language Association, 1995).

[65] David C. GREETHAM, *Theories of the Text* (Oxford: Oxford University Press, 1999).

[66] Louis HAY and Péter NAGY (eds), *Avant-texte, texte, après-texte* (Paris: Éditions du CNRS, 1982).

[67] Helmut HILLER and Wolfgang STRAUSS (eds), *Der deutsche Buchhandel. Wesen — Gestalt — Aufgabe*, 2nd ed. (Gütersloh: Bertelsmann, 1962).

[68] Achim HÖLTER, "Tiecks Bibliothek", in *Ludwig Tieck. Leben — Werk — Wirkung*, ed. by Claudia Stockinger and Stefan Scherer (Berlin: De Gruyter, 2011).

[69] Max HORKHEIMER and Theodor W. ADORNO, *Dialektik der Aufklärung. Philosophische Fragmente* (Frankfurt/Main: Fischer, 1969).

[70] Axel HORSTMANN, "'Erkenntnis des Erkannten'. Philologie und Philosophie bei August Boeckh", *Zeitschrift für Germanistik*, 20.1 (2010), pp. 64-78.

[71] Matthew JOCKERS, *Macroanalysis. Digital Methods and Literary History* (Chicago: University of Illinois Press, 2013).

[72] Helmuth KIESEL and Paul MÜNCH, *Gesellschaft und Literatur im 18. Jahrhundert. Voraussetzungen und Entstehung der literarischen Markts in Deutschland* (Munich: Beck, 1977).

[73] Birgit KUHBANDNER, *Unternehmer zwischen Markt und Moderne. Verleger und die zeitgenössische Literatur an der Schwelle zum 20. Jahrhundert* (Wiesbaden: Harrassowitz, 2008).

[74] Stefan LANGE and Tilman SANTARIUS, *Smarte grüne Welt? Digitalisierung zwischen Überwachung, Konsum und Nachhaltigkeit* (Munich: oekom, 2018).

[75] James LEGG, "Chapter 14: Underground Bookstore and Old Bodleian Access Project", in *Transforming the Bodleian*, ed. by Michael Heaney and Catriona Jeanne Cannon (Berlin: De Gruyter, 2013), pp. 191-209, https://doi.org/10.1515/9783110289398.

[76] Mark LEHMSTEDT, "'Ich bin nun vollends zur Kaufmannsfrau verdorben'. Zur Rolle der Frau in der Geschichte des Buchwesens am Beispiel von Friedrike Helene Unger (1751–1813)", *Leipziger Jahrbuch für Buchgeschichte*, 6 (1996), pp. 81-154.

[77] John B. LYON, "'Anzeigen', Supplementarity, and Modernity in Fontane", in *Fontanes Medien*, ed. by Peer Trilcke (Berlin: De Gruyter, 2022), pp. 375-390.

[78] Marshall MAC LUHAN, *the gutenberg galaxy - the making of typographic man* (Toronto: University of Toronto Press, 1962).

[79] Marshall MAC LUHAN, *Understanding Media: The Extensions of Man* (1964) (Cambridge, MA: MIT Press, 1994).

[80] Alberto MANGUEL, *A History of Reading* (Toronto: Vintage Canada Edition, 1998).

[81] Kevin MARQUET, Jacques COMBAZ and Françoise BERTHOUD, "Introduction aux impacts environnementaux du numérique", *1024, bulletin de la Société Informatique de France* (2019), pp. 85-97.

[82] Franco MORETTI, *Distant Reading* (London/New York: Verso, 2013).

[83] François MOUREAU, *La plume et le plomb. Espaces de l'imprimé et du manuscrit au siècle des Lumières* (Paris: PUPS, 2006).

[84] Brigitte OUVRY-VIAL, "Reading Seen as Commons", *Participations. Journal of Audience and Reception Studies*, 16.1 (May 2019), https://www.participations.org/16-01-09-ouvry.pdf.

[85] Andrea PATAKI-HUNDT, "Bestandserhaltung", in *Handbuch Archiv*, ed. by Marcel Lepper and Ulrich Raulff (Stuttgart: Metzler, 2016), pp. 218-224.

[86] Roger PAULIN, *Ludwig Tieck. Eine literarische Biographie* (Munich: Beck, 1988).

[87] Roger T. PÉDAUQUE, *Document et modernités*, 2006, http://archivesic.ccsd.cnrs.fr/sic_00001741.

[88] Panu PIHKALA, "Eco-anxiety", in *Situating Sustainability. A Handbook of Contexts and Concepts*, ed. by C. Parker Krieg and Reetta Toivanen (Helsinki: Helsinki University Press, 2021), `https://www.jstor.org/stable/j.ctv26qjj7d.14`.

[89] Andrew PIPER, *Book was There. Reading in Electronic Times* (Chicago: University of Chicago Press, 2012).

[90] Paul RICOEUR, *Temps et Récit III. Le temps raconté* (Paris: Seuil, 1985).

[91] Paul RICOEUR, *Du texte à l'action* (Paris: Seuil, 1986).

[92] Eva RIETZSCHEL (ed.), *Gelehrsamkeit ein Handwerk? Bücherschreiben ein Gewerbe? Dokumente zum Verhältnis von Schriftsteller und Verleger im 18. Jahrhundert in Deutschland* (Leipzig: Reclam, 1982).

[93] Torsten ROEDER, "Review of Juxta Web Service, LERA, and Variance Viewer. Web Based Collation Tools for TEI", *RIDE - A Review Journal for Digital Editions and Resources*, 11 (2020), `https://doi.org/10.18716/ride.a.11.5.`.

[94] Theodore R. SCHELLENBERG, *The Appraisal of Modern Public Records* (Washington: Bulletin of the National Archives 8, 1956).

[95] Siegfried SCHMIDT, *Die Selbstorganisation des Sozialsystems Literatur im 18. Jahrhundert* (Frankfurt/Main: Suhrkamp, 1989).

[96] Emil STAIGER (ed.), *Der Briefwechsel zwischen Schiller und Goethe* (Frankfurt/Main: Insel, 1966).

[97] Jochen STROBEL, "Dresden, Berlin und Potsdam", in *Ludwig Tieck. Leben — Werk — Wirkung*, ed. by Claudia Stockinger and Stefan Scherer (Berlin: De Gruyter, 2011), pp. 108-119.

[98] Peter SUBER, *Open Access* (Cambridge, MA/London: MIT Press, 2012), `https://mitpress.mit.edu/books/open-access`.

[99] Denis THOUARD and Christian BERNER (eds), *L'interprétation. Un dictionnaire philosophique* (Paris: Vrin, 2015).

[100] Ciaran B. TRACE, "Beyond the Magic to the Mechanism: Computers, Materiality, and What It Means for Records to be 'Born Digital'", *Archivaria*, 72 (Fall 2011), pp. 5-27.

[101] Jutta WEBER (dir.), *Sternstunden eines Mäzens: Briefe von Galilei bis Einstein aus der Sammlung Ludwig Darmstaedter* [exhibition catalogue] (Berlin: Staatsbibliothek zu Berlin, 2008).

[102] Rob WILKIE, *The Digital Condition* (New York: Fordham University Press, 2011).

[103] Heinz WITTENBRINK, "Zwischen Resilienz und Regeneration", in *Playbook Klimakultur. Strategien für einen nachhaltigen Kulturwandel*, ed. by Birgit Lurz, Wolfgang Schlag and Thomas Wolkinger (Wien: Bundesministerium Europäische und Internationale Angelegenheiten, 2021), pp. 145-150, `https://www.bmeia.gv.at/fileadmin/user_upload/Zentrale/Kultur/Publikationen/Playbook_Klimakultur_Web_20210910.pdf`.

[104] Reinhard WITTMANN, *Geschichte des deutschen Buchhandels*, 3rd ed. (Munich: Beck, 2011).

[105] Edda ZIEGLER, *Julius Campe. Der Verleger Heinrich Heines* (Hamburg: Hoffmann und Campe, 1976).

[106] Edda ZIEGLER, *Buchfrauen. Frauen in der Geschichte des deutschen Buchhandels* (Göttingen: Wallstein, 2014).

About the Team

Alessandra Tosi was the managing editor for this book.

Elizabeth Rankin and James Hobson performed the copy-editing and proofreading.

Jeevanjot Kaur Nagpal designed the cover. The cover was produced in InDesign using the Fontin font.

Gilles Blanchard typeset the book in LaTeX, and Anne Baillot indexed the manuscript.

Cameron Craig produced the paperback and hardback editions.

This book has been anonymously peer-reviewed by experts in their field. We thank them for their invaluable help.

www.ingramcontent.com/pod-product-compliance
Lightning Source LLC
Chambersburg PA
CBHW071215240726
48654CB00009B/801